Physical Geology

INTRODUCING GEOLOGY SERIES

Fossils
by F. A. MIDDLEMISS

British Stratigraphy
(Revised Metric Edition)
by F. A. MIDDLEMISS

Rocks and Minerals
by JANET WATSON

Physical Geology
by J. R. L. ALLEN

INTRODUCING GEOLOGY: 3
EDITOR: J. A. G. THOMAS
*Head of Geography and Geology Department,
Verdin Comprehensive School, Winsford, Cheshire*

Physical Geology

JOHN R. L. ALLEN
Professor of Geology, University of Reading

London

A THOMAS MURBY PUBLICATION OF
GEORGE ALLEN & UNWIN LTD

First published in 1975

This book is copyright under the Berne Convention. All rights are reserved. Apart from any fair dealing for the purpose of private study, research, criticism or review, as permitted under the Copyright Act, 1956, no part of this publication may be reproduced, stored in a retrieval system, or transmitted, in any form or by any means, electronic, electrical, chemical, mechanical, optical, photocopying, recording or otherwise, without the prior permission of the copyright owner. Inquiries should be addressed to the publishers.

© George Allen & Unwin Ltd 1975

ISBN 0 04 550022 3

Printed in Great Britain
in 11 point Times New Roman
by Cox & Wyman Ltd, London, Fakenham and Reading

Contents

Preface	8
Acknowledgements	9
1 The Face of the Earth	11
2 Weathering and Soils	17
3 Entrainment, Transport and Deposition of Sediment	25
4 The Work of Rivers and Underground Waters	37
5 The Work of the Wind	52
6 The Work of the Sea: Cliffed and Sandy Coasts	62
7 The Work of the Sea: Continental Shelves	74
8 The Work of the Sea: the Ocean Deeps	85
9 The Work of Ice	100
10 The Restless Lithosphere: Evidence from the Past	112
11 The Restless Lithosphere: Evidence of Movement at the Present Day and in the Geologically Recent Past	126
Index	139

Preface

Much of the evidence concerning the history of the Earth is preserved in the stratified rocks so prevalent in the crust. Before we can appreciate this history we must learn something about the formation of sediments at the Earth's surface and about the internal movements within the Earth which affect sedimentation and which influence the already existing stratified deposits. This book provides an introduction to the origin of sediment, the chief physical agents of sedimentation and their products, and to earth-movements and their effects on sedimentation and strata. It is intended to be used by candidates studying for the GCE ordinary level and similar examinations. I have covered as far as possible the topics in physical geology at this level cited in the various syllabuses, the few omissions being dictated partly by limitations of space but mainly by recent reorientations in the subject. In the past decade or so, physical goeology has become markedly better founded in mechanical principles, while the study of the marine environment has advanced enormously to a position representing a substantially correct balance. These developments are given due weight. The teacher will also find suggestions for simple, inexpensive experiments which can be done to develop insight into many of the points I discuss. Many more will no doubt come to mind, in view of the nature of the subject (not entirely descriptive). There are of course innumerable field projects, by river and sea-shore, for example, which are latent in my treatment and which could profitably be pursued.

I am greatly indebted to Mr J. A. G. Thomas (Head of Geography and Geology, Verdin Comprehensive School, Winsford, Cheshire), who carefully criticised the whole of this book in manuscript, and also to Mrs G. Raistrick, Dr A. Parker and Mr J. Towsey, who commented most helpfully on portions of the work. I should also like to thank Mrs B. Tracey for typing the manuscript and Mr J. L. Watkins for a substantial part of the photographic work.

J.R.L.A.

Acknowledgements

The following individuals and organisations are warmly thanked for very kindly allowing me to reproduce photographs taken in the course of their activities.

Figures 2.1, 2.3, 3.7, 5.8, 6.6, 9.2B, 9.2C, 9.2D, 9.7, 10.3, 10.4, 10.14, 10.15, 10.16. Reproduced by permission of the Controller, Her Majesty's Stationery Office. Crown Copyright Geological Survey photographs.

Figure 3.2. Dr O. C. Lloyd, Department of Pathology, University of Bristol and the University of Bristol Spelaeological Society. Reproduced from *The Caves of Northwest Clare, Ireland*. Copyright reserved.

Figures 3.5, 3.6. Dr P. Worsley, Department of Geography, University of Reading.

Figure 4.7. Dr B. J. Bluck, Department of Geology, University of Glasgow.

Figures 4.8, 5.3, 5.10, 6.5. Aerofilms Limited, London. Copyright reserved.

Figure 4.10. Dr G. E. Williams, St Peters, South Australia 5069.

Figures 5.4, 5.5. Dr K. W. Glennie, Shell Research B.V., and Elsevier Publishing Company, Amsterdam. Reproduced from K. W. Glennie, *Desert Sedimentary Environments*. Copyright reserved.

Figures 7.3, 7.9, 7.10. Dr M. R. Dobson, Department of Geology, University College of Wales, Aberystwyth.

Figures 8.11A, 8.11B, 8.11C, 8.11D, 8.11E, 8.11F. Dr A. S. Laughton, National Institute of Oceanography, Wormley, Godalming. Copyright reserved.

Figures 9.2A, 11.2. Director, Surveys and Mapping Branch, Ministry of Energy, Mines and Resources. Canadian Government Copyright.

Figure 9.9. Dr R. B. Pelletier, Bedford Institute of Oceanography, Dartmouth, N.S., Canada.

Figure 9.10. Dr L. A. Frakes, Department of Geology, Florida State University. Reproduced from the *Bulletin of the Geological Society of America*, v. 82, p. 1593 (1971).

Chapter 1
The Face of the Earth

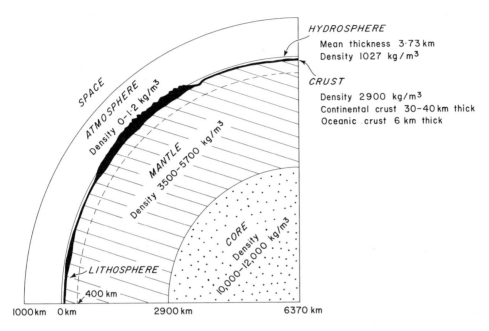

Figure 1.1 The layered structure of the Earth. For clarity the thickness of the crust is exaggerated.

Introduction

Our Earth is a very slightly flattened *spheroid* with a *layered internal structure* (Figure 1.1). The radius at the poles is 6356km, a little less than the radius of 6378km measured at the equator. The layered structure is known from the studies that geophysicists have made of the velocities and paths of earthquake and other powerful shock waves travelling within the Earth.

The central layer, known as the *core*, is very dense and largely liquid, consisting it is thought of a mixture of nickel and iron. The core extends outwards to within about 2900km of the Earth's surface. The next and by far the most voluminous layer is the solid *mantle*. It is much less dense than the core, and is believed to consist of silicate minerals rich in iron and magnesium, such as olivine and pyroxene. The mantle extends from the edge of the core outwards to the *Mohorovičić* or M *discontinuity*, a level at which both the velocity of earthquake waves and the density of the Earth's interior decrease significantly. The M discontinuity varies in level, from a maximum depth of about 60km to as shallow as 10km. Finally, between the M discontinuity and the Earth's surface, we have the outermost layer, the *crust*. The crust, like the mantle beneath it, consists mainly of silicate minerals, but species rich in alumina, such as the feldspars, are now plentiful. The crust beneath the oceans, however, differs in composition and density from the crust forming the continents. The upper part of the mantle forms with the crust a special layer called the *lithosphere*. The materials which form the lithosphere are more rigid than occur in the remaining mantle below.

The physical geologist recognises that the Earth possesses two more layers of importance (Figure 1.1). The water in the ocean basins forms the *hydrosphere*, covering 71 per cent of the Earth's surface. The hydrosphere is a discontinuous, incomplete layer of very variable but relatively very small thickness, the average being 3.73km. The outermost layer of all is the *atmosphere* which forms man's natural environment and that of many other organisms. It is a continuous shell, wrapping around the whole Earth, but the thickness is difficult to define on account of the fact that the density of air lessens with increasing height above the ground. At sea-level, for example, the density of the air is about 1.2kg per cubic metre, but at a height of several hundred kilometres we find the nearly perfect vacuum of space.

Thus the Earth consists of a series of layers, differing from each other in state and composition and decreasing in density outwards. Ignoring the very dense but liquid core, it would seem that the planet Earth presents an ordered sequence of layers: the solid mantle and crust are overlain by the liquid hydrosphere, in turn enveloped by the gaseous atmosphere.

Movements within the Earth and the geocycle
Although we spoke of a solid mantle and a solid crust, none of the layers of the Earth is perfectly rigid when viewed on a geological time-scale. The hydrosphere and atmosphere, and a large part of the core, are made of substances classed as fluids. Everyone is familiar with the fact that fluids, such as ocean water and the air, change shape when acted upon by suitable forces. They become *deformed*. The mantle and crust can also behave in a fluid-like way, provided they have enough time. In this respect they resemble pitch or toffee, which *flows* when acted upon gradually enough by forces. When the forces act quickly, however, toffee and pitch *shatter*, behaving in a way we ordinarily associate with solids. As you will discover later, both the mantle and crust are moving very slowly under the action of the various forces that operate inside the Earth. When we reflect that the atmosphere and hydrosphere also are in constant motion, the Earth emerges as filled with potential for change.

The fact that the layers we have discussed are *mobile* is profoundly important in physical geology. The slow flow of the mantle seems to cause huge plate-like pieces of the lithosphere to *slide horizontally* over the Earth's surface, thus in time bringing geologically unrelated rocks together. *Vertical* as well as horizontal *crustal movements* are also possible. Thus parts of the crust differing in thickness appear able to 'float' in equilibrium upon the more dense mantle. Any change in the thickness or density of a part is followed by a readjustment of its vertical position. We are, of course, hardly aware of movements of these kinds, because they normally occur so very slowly. But the mobility of the hydrosphere and atmosphere is an everyday experience:

rivers flow, waves beat on the shore, and the wind blows around us. We also notice that these *agents*, because of their movement, *transport* mineral particles broken from crustal rocks. As the result, fresh rock is exposed to *weathering*, and elsewhere those particles are *deposited* to form a new layer of sediment which one day could become rock. Multiply these actions a billion-fold, and it is easy to see how in the course of geological time, whole mountain ranges can be worn away and large crustal depressions filled up. Indeed, the rapid weathering of some portion of the crust, or the infilling of some major basin, may be promoted by appropriate movements in the Earth's interior.

Thus the history of the Earth is the record of the continuing *dynamic interactions* between the mantle, crust, hydrosphere and atmosphere. These interactions can be described in terms of a model, the *geocycle* (Figure 1.2).

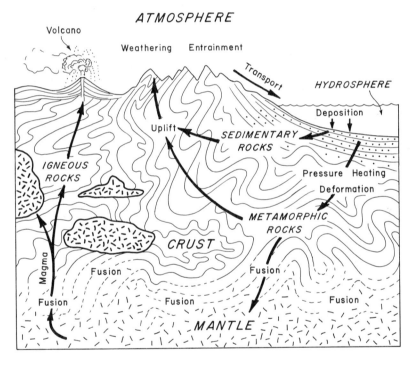

Figure 1.2 A model of the geocycle.

They principally express themselves by changes at one particular world-wide surface, namely, that between the crust on the one hand and the hydrosphere and atmosphere on the other. It may be figuratively called the *face of the Earth*. You can see how, in the course of time, materials pass through the face of the Earth into the hydrosphere or atmosphere, and then back again into the Earth's interior.

Classifications of the face of the Earth

The most obvious classification of the face of the Earth is between *land* and *sea*, though this not necessarily the most useful or important (Figure 1.3). In this essentially geographical classification, we recognise that the land forms the more elevated parts of the crust, whereas the sea infills the hollows of the *ocean basins*. The ocean basins are formed of (a) the *mid-ocean ridges*, which are submerged mountain systems; (b) the *ocean basin floors*; and (c) the *continental margins*. The land areas are divided into (a) *low-lying plains*; (b) *mountain ranges and plateaux*.

A second classification may be based on the geophysical character of the crust beneath the face of the Earth (Figure 1.3). The crust beneath the land

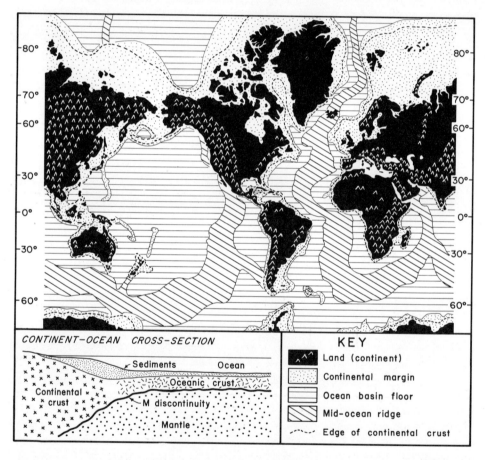

Figure 1.3 Geographical and geophysical classifications of the face of the Earth.

areas and the continental margins is relatively thick and consists of *granite-like rocks*. This is the *continental crust*. On the other hand, the ocean basin floors and the mid-ocean ridges are underlain by a thinner and denser type of crust, the *oceanic crust*, composed of *basalts*.

The third classification (Figure 1.4) we shall consider is based on geological

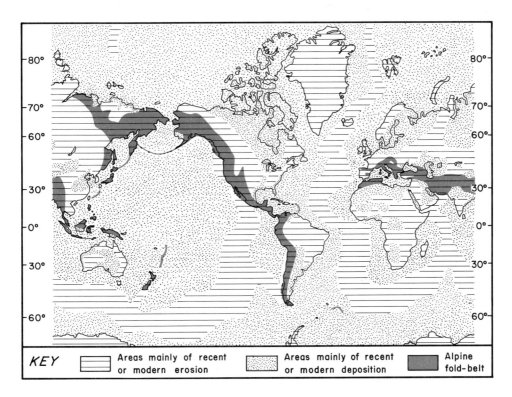

Figure 1.4 Classification of the face of the Earth using geological criteria.

points of view and introduces many of our later discussions on sedimentary environments and tectonics.

Over large parts of the face of the Earth, the balance of processes leads to the *weathering and destruction* (erosion) of the crustal rocks. On the land the areas being so denuded include geologically young mountain chains (e.g. Alpine folds) and a number of high-level plateaux formed of somewhat older rocks. Debris from these regions is carried to the sea by many large rivers. Comparatively few of these empty into the Pacific Ocean, the largest of the oceans, whereas the Arctic, Indian and Atlantic Oceans receive many of the largest rivers. Also, parts of the face of the Earth beneath the sea are erosional in character. In many places, such as large areas in the English Channel, the water is so shallow and swept by such strong currents that no loose sediment can accumulate on the sea bed and hard rocks are laid bare. In addition, erosion occurs over the more rugged parts of the mid-ocean ridges, where basalt lavas are exposed on the sea floor. In many other places in the oceans bottom currents have grown strong enough to prevent the deposition of the fine suspended sediment. The earlier deposits then become scoured away, and often channelled or grooved, as the result of the action of currents.

On the other parts of the face of the Earth new layers of sediment are being formed because *deposition* prevails. These depositional areas are very varied in character. Some lie well inland and have an internal or partly internal drainage; for example, the great deserts of Central Asia, and the river floodplains and swamps of the River Paraguay in South America. Many other important areas of deposition lie along the edges of the land. These include the plains which border the Gulf of Mexico, where river and coastal sediments are closely associated over a distance of nearly 1500km, and the sandy barrier islands and intertidal mud flats represented by the Friesian Islands and the Wadden Sea on the eastern shores of the North Sea. The largest depositional regions, however, are found on the ocean basin floors, especially the huge smooth areas known as the *abyssal plains* and the floors of the long narrow depressions called *deep-sea trenches*. There are also areas of deposition on the continental margins and some on the smoother parts of the mid-ocean ridges. As you can see from the geocycle (Figure 1.2), the ocean basins are the *sinks* to which are transported the sediments liberated through weathering from land areas, which are the *sources*.

There is a strong association between the belts of Alpine (Tertiary) folding —perhaps the youngest erosional parts of the face of the Earth—and many of the largest depositional areas (Figure 1.4). Thus many such areas are adjacent to and parallel with the Alpine fold-belts. For instance, west of the Andes and the mountains of Central America lie deep-sea trenches richly supplied with sediment. There is another major depositional region immediately to the east of the Andes, but it is filled with river and swamp deposits. A contrasting situation is provided by the Central Asian parts of the fold-belt. In this region the mountain chain is bordered to the north by huge basins of interior drainage containing deserts. However, to the south of the Himalayan part of the fold-belt, lie the enormous and well-watered plains of the Indus and Ganges rivers with their thick sediment. Yet another contrast can be found in the East Indies, in the arc of mountainous islands dominated by Sumatra and Java. South of the arc lie deep-sea trenches, but on the north is a shallow sea fringed by brackish swamps.

Several major depositional regions have no apparent association with Alpine folding. Examples are the deserts of southern Africa and Australia, and the extensive river floodplains and swamps of the Ob and Yenisei of Siberia.

Most of the remainder of this book will consist of a more detailed study of points that have been raised in discussing the geocycle and the different classifications that may be proposed for the face of the Earth. We shall discuss the way rocks weather, the erosion and transport of weathered materials, the environments in which sediments become deposited and, finally, the interplay between earth-movements and sedimentation.

Chapter 2
Weathering and Soils

Introduction

Weathering is the disintegration and decomposition *in situ* of the rocks exposed on the face of the Earth. Some weathering processes are *mechanical*, whereas others are *chemical*. Others strictly are *biochemical* or *biomechanical*, because they involve the life-activities of organisms. Weathering is most obvious on the land. Here we can see rocks being slowly changed into a blanket of loose *soil* in which plants, some of which serve as food, will grow. But weathering also causes the decay of rocks exposed on the bed of the seas and oceans, as well as those on the bottom of rivers and lakes. Everywhere the rocks at the Earth's face are decaying and loosening, in preparation for transport to new sites, where fresh layers of sediment will be formed and finally changed into rock.

Mechanical weathering

The mechanical processes of weathering involve the *forcing apart* of the rock,

Figure 2.1 Frost-riven quartzite at an elevation of approximately 1100m near Ben Nevis, Grampian Mountains, Scotland.

along the surfaces of weakness dividing it. On the largest scale, the rock may be split by joints and bedding planes several decimetres or metres apart. Surfaces of cleavage and lamination may part rocks on an intermediate scale. On the smallest scale, the required weaknesses are the boundaries between the individual crystals or mineral grains in the rock.

A powerful expanding force is exerted by crystals growing from a liquid. Hence substances that crystallise along surfaces of weakness in rocks can act like a wedge, and shatter or split the rock. The most important case of this process of mechanical weathering is ice-crystal growth, called *ice-wedging* or *frost-riving*, which results in coarse angular debris (Figure 2.1). It occurs only in those regions of the Earth where the temperature oscillates about the freezing point of water, namely, in high and polar latitudes and in mountainous areas. The landscape that develops due to the action of this process is marked by bold relief and the presence of sharp jagged ridges and peaks. In hot dry regions, however, rocks weather mechanically partly through the growth of crystals of salts in cracks and fractures. These salts—themselves the products of chemical weathering—accumulate because in these places evaporation exceeds precipitation and soluble materials cannot be carried away. Mechanical weathering by crystal growth may operate at the sea coast, where in dry weather salt water is evaporated on the surfaces of the rocks.

Plants also exert powerful wedging forces during growth (Figure 2.2). Where soils are thin, grasses and trees anchor themselves in the rock by

Figure 2.2 The growth of trees and their roots disrupts the Roman wall at Silchester (*Calleva Atrebatum*), Hampshire.

squeezing their roots along joints and bedding, thus separating adjacent pieces. You will probably have noticed the way the roots of trees growing on a ruined building force the stones apart and eventually send them tumbling to the ground.

It is often said that rocks will disintegrate if they experience a sufficiently large daily *heating and cooling*, such as happens in hot deserts, where the day and night temperatures of the ground surface may differ by as much as 50 °C. This is thought to occur because the rocks heat up or cool unevenly and because the different mineral particles in the rocks expand or contract at contrasted rates. Laboratory experiments do not entirely support this claim, but the problem remains open because of the long periods and complex circumstances under which the natural processes have operated. There is some evidence that rocks break up due to heating and cooling if they are also being weakened chemically.

Chemical weathering
Chemical weathering depends largely on two facts. Firstly, no mineral is completely insoluble in natural water or entirely inert chemically. Secondly, water is a very reactive substance despite appearances to the contrary. Therefore chemical weathering is ubiquitous, because water is present almost everywhere in the rocks near the surface, either as a pore-filling fluid or as a film of moisture coating each grain. Rocks and soils even in hot deserts are regularly moistened, by the dew that forms in the cold of night.

The *chemical instability* of the common rock-forming minerals in the presence of water is readily demonstrated. For example, a strongly acid reaction to a suitable test paper will be obtained if fresh pyrites, FeS_2, is ground with water in a mortar. But the reaction will prove alkaline on using olivine, $(MgFe)_2SiO_4$. The common minerals can in fact be arranged in a series of decreasing resistance to chemical weathering, the *mineral stability series*. From the least to the most reactive, they are: kaolinite (the clay mineral of china clay), quartz, orthoclase (potassium feldspar), albite (sodium plagioclase), anorthite (calcium plagioclase), hornblende, augite, and olivine. Calcite and dolomite, the two commonest carbonate minerals, also have little resistance to chemical weathering.

The reactions that involve water and the common minerals are very complicated in their details, but can be classified under the four headings of *hydrolysis*, *oxidation*, *carbonation*, and *hydration*. The laws of *chemical equilibrium* are obeyed by these weathering reactions, just as by the more familiar reactions made in the laboratory. The soluble products of hydrolysis, for example, must be constantly removed, or the weathering will stop. Hence chemical weathering is promoted by leaching. It should therefore be most effective in regions where precipitation, preferably high, exceeds evaporation and where temperatures are high. The rocks of the humid tropics are for this reason more deeply and completely weathered chemically than those of cold or arid regions. Finally, there must always be a sufficient supply of anions with which the cations can combine.

Hydrolysis is the most important of the weathering reactions, and it affects chiefly the feldspar and ferromagnesian minerals. During this reaction, metallic ions in the mineral are exchanged for hydrogen ions in the water. The hydrolysis of potassium feldspar, for example, may be represented by the reaction

$$\underset{\text{feldspar}}{4KAlSi_3O_8} + \underset{\text{water}}{6H_2O} \rightarrow \underset{\text{kaolinite}}{Al_4[Si_4O_{10}](OH)_8} + \underset{\text{silica}}{8SiO_2} + \underset{\text{potash}}{4KOH}$$

The *potash* and commonly the *silica* wash away in solution and the insoluble *kaolinite* is left behind at the place of weathering. Another clay mineral, *illite*, may form instead of kaolinite if the potash is not all leached.

The hydrolysis of the ferromagnesian minerals commonly involves oxygen, as may be illustrated by the weathering of *olivine*, the least stable of this kind of mineral

$$\text{olivine} + \text{water} + \text{oxygen} \rightarrow \text{serpentine} + \text{haematite} + \text{silica}.$$

The substances formed by the weathering of ferromagnesian minerals are a little bulkier than their parents. Hence when rocks rich in these minerals begin to weather chemically, a tendency for expansion may become apparent. Since weathering proceeds inwards from joints, the expansion is often expressed as shell-like partings, giving the familiar *onion-skin* or *spheroidal weathering* of

Figure 2.3 The igneous rock in this outcrop is rich in ferromagnesian minerals and is weathering spheroidally.

basic igneous rocks (Figure 2.3).

Chemical weathering involving *oxidation* and *hydration* may be illustrated by the reaction between pyrites and oxygen in the presence of water. In detail the reaction is certainly very complex, but it may be represented by

$$4FeS_2 + 8H_2O + 15O_2 \rightarrow 2Fe_2O_3 + 8H_2SO_4$$
$$\text{pyrites} \quad \text{water} \quad \text{oxygen} \quad \text{limonite} \quad \text{sulphuric acid}$$

The sulphuric acid produced may combine with unweathered pyrites to yield native *sulphur*, as follows

$$FeS_2 + H_2SO_4 \rightarrow FeSO_4 + H_2S + S$$
$$\text{pyrites} \quad \text{sulphuric acid} \quad \text{ferrous sulphate} \quad \text{hydrogen sulphide} \quad \text{sulphur}$$

A visit to a colliery spoil-heap or to coastal cliffs of clay will provide many illustrations of these reactions. *Limonite*, $2Fe_2O_3.H_2O$, and native sulphur can all be found on the weathered shale, and the groundwater is likely to prove strongly acid. *Gypsum*, $CaSO_4$, in the form of needle crystals can usually be discovered as well. It forms by the reaction of sulphuric acid with any calcite, $CaCO_3$, present in the rocks.

Carbonation is the weathering reaction that principally effects the carbonate minerals in limestones, of which calcite and dolomite are the most important. Water and carbon dioxide, dissolved from the atmosphere, form a mixture which proceeds to react with, for instance, dolomite as follows

$$CaMg(CO_3)_2 + 2H_2O + 2CO_2 \rightarrow Ca(HCO_3)_2 + Mg(HCO_3)_2$$
$$\text{dolomite} \quad \text{water} \quad \text{carbon dioxide} \quad \text{calcium bicarbonate} \quad \text{magnesium bicarbonate}$$

These *bicarbonates* are highly soluble and can easily be leached away by rain. The solutional features found on limestone outcrops and on the walls of limestone caves, all prove the corrosive power of water containing dissolved carbon dioxide.

Chemical weathering is helped by the substances produced by living plants, for example, the acids secreted by *lichens* colonising rock surfaces. It is also assisted by the *humic acids* formed during decay of plant litter.

Soils and their distribution

The various weathering processes we have considered, combined with the activities of plants and animals, and the effects of their remains, cause the surface rocks to undergo a slow chemical and physical degradation into *soil*. Although very variable in character, soils can be arranged between a small number of natural *classes*, whose distribution is commonly world-wide and related to climate, especially temperature and precipitation, more than to parent rock. These major soil classes often have names which derive from the Russian, because it was in Russia that many of the most important early studies of soils were made.

Each soil class can be recognised by its characteristic *vertical profile*, or the sequence of distinctive horizontal layers, called *soil horizons*, which forms with increased weathering of the parent rock. The way in which to expose

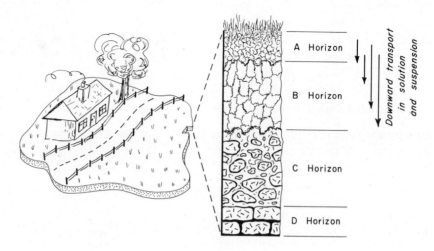

Figure 2.4 A model of the soil profile.

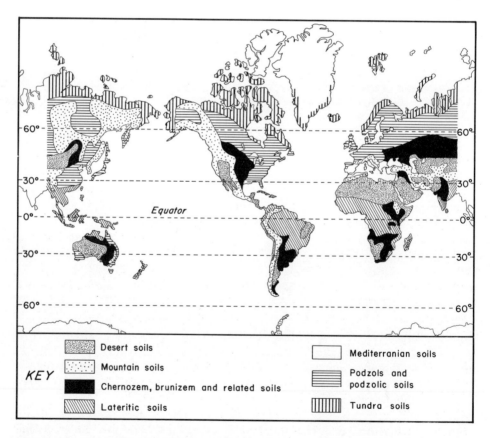

Figure 2.5 A greatly simplified world soil map.

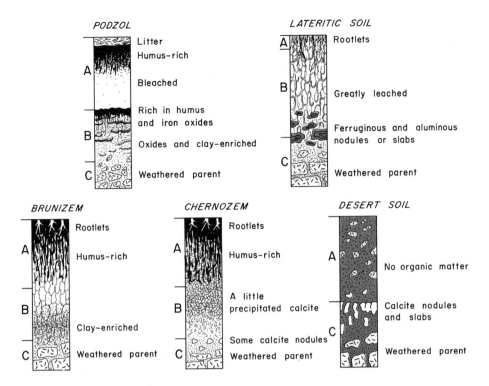

Figure 2.6 Some simplified soil profiles.

these profiles is to dig a shallow inspection pit in the earth, keeping the sides vertical and smooth. The soil horizons will be found to differ from each other in *composition, colour, texture* and *structure* (Figure 2.4). Proceeding downwards from the surface to the unaltered parent rock, they are named the A, B, C and D horizons. The *A horizon* is the zone of *eluviation* or leaching. Water percolating downwards through this horizon has removed from it all the products of weathering that could be transported in solution or suspension. The *B horizon* is the zone of *illuviation*, or accumulation, of some or all of the substances carried down from the A horizon. The *C horizon* consists of partly disintegrated and somewhat altered parent rock. It is commonly known as the *subsoil*. The *unaltered rock* constitutes the *D horizon*. Figure 2.5 is a map showing the distribution of the major soil classes. Only those that are most important and interesting can be discussed in detail.

Although *podzols* (Figure 2.6) are most widely distributed in cool, humid parts of the world, they are not uncommon in the subtropics. Podzols are strongly acid. The A horizon consists of several subordinate layers. At the top is undecomposed plant litter, which grades downwards into a brown layer of partly decomposed organic material. Below this is a black horizon rich in humus. The lowest layer of the A horizon is a grey to white deposit which has lost through leaching its humus, clay, iron oxide and alumina. The B horizon has an upper dark brown to black layer of precipitated humus, overlying a grey, yellow or red layer rich in iron oxides, alumina and clay. Podzols are common in many parts of Britain, in the low-lying east and south as well as in the mountainous west and north.

The *lateritic soils* (Figure 2.6) are also acid and strongly leached. They occur in the tropics and subtropics where there is high rainfall, intense leaching of the soil, and an oxidising warm environment. The soil profile is often many metres deep, and yellow to red colours predominate. The accumulation of organic matter is inhibited by the oxidising conditions, and the high rainfall ensures that soluble weathering products are swiftly removed. Hence lateritic soils are strongly enriched with iron and aluminium oxides and hydroxides. The leaching is so intense that even the relatively insoluble silica is almost entirely removed in solution. Lateritic soils are often quarried as a building material, or as a source of aluminium ore.

Brunizems (Figure 2.6) occur where precipitation is moderately in excess of evaporation. Hence these soils are mildly acid to neutral, but with fewer and less obvious layers than podzols. The A horizon is organic-rich and coloured dark brown. In the B horizon we find an upper light coloured layer, with below a yellow to red layer rich in clay and other materials leached from above.

Chernozem or black earth (Figure 2.6) is an important and widely distributed type of soil that was first described from the Russian steppes, which are level treeless plains. The A horizon is usually several decimetres thick, dark brown to black in colour, and very mildly acid. It contains an abundance of organic material. The B horizon is yellow to brown and mildly alkaline. At its base are to be found *calcium carbonate nodules* that have been precipitated in the soil as the result of the leaching and accumulation processes. These concretions are known as *calcretes*, and they are often so abundant as to form a continuous thin limestone bed within the soil. Their position below the surface varies with the amount of rainfall. In the wetter regions the nodules are sparse and occur at a depth of 2m or more, but in drier areas, bordering the desert soils, are plentiful and lie only 20 to 30cm down.

Desert soils (Figure 2.6) are shallow and rich in undecomposed rock debris. They are poor or lacking in organic material, and the dominance of evaporation over precipitation ensures that leaching is relatively less effective than elsewhere. Generally speaking, layering is crude or not developed. Calcrete is widespread at shallow depths in desert soils, and gypsum and sodium sulphate may also be encountered. Desert soils vary in colour from grey in cool climates to yellow or red in hot ones. They are generally alkaline. The strongest colours seem to require a higher temperature and a smaller precipitation for their formation than the pale ones.

Chapter 3
Entrainment, Transport and Deposition of Sediment

Introduction
Products of weathering stay where they are formed unless some natural *agent* picks them up and carries them away. The picking up of such products is called *entrainment*, and is due to the operation of *processes of entrainment*. *Transport* and *transport processes* refer to the carrying of products of weathering from one place to another. *Deposition* through the operation of *depositional processes* occurs when a transporting agent lays down weathered material earlier picked up. The terms erosion and denudation are often used in regard to entrainment and transport but have no generally accepted rigorous meaning.

Entrainment, transport and deposition are crucial factors in physical geology. Without them soil-formation would cease, the landscape would no longer change, and deposition of sediment (weathering products) on the edges of the land and in the seas and oceans would virtually stop. The fact that we can recognise these processes today means that the Earth has continued to change right up to the present. The *rates of change* are very small in terms of a human lifetime, but on a geological time-scale represent significant effects.

Agents involved
The agents involved in entrainment, transport and deposition are (a) *rivers* and *overland sheet flows*, (b) *underground streams*, (c) *groundwater*, (d) *waves*, (e) *tides*, (f) *turbidity currents*, (g) *oceanic circulations*, (h) *winds*, (i) *glaciers*, and (j) *flowing soils*. They work in different ways and at different rates, some at sea and others on the land (Figure 3.1). Some that operate on the land, such as glaciers, are restricted to particular climates.

Rivers
Rivers and overland sheet flows (thin layers of flood water) are the most important agents affecting the land. They pick up sediment through *hydraulic* and *mechanical activities*. If you have ever paddled in a river, you will know that the current exerts a push or *pressure force* against any object in its path. The flow also exerts on the object a drag or *shear force*, which is the kind of force arising when a block of wood is slid over the flat top of a table. If these forces are large enough compared with the weight of a stationary sediment particle on the stream bed, that particle is entrained by the action of the river's hydraulic forces. The particle can also be entrained mechanically, if it is struck hard enough by an already moving particle. The sediment picked up in these ways is transported by one of two different methods. Sand and gravel particles, which are relatively large and heavy, travel close to the bed as a *bed load*, *sliding* or *rolling* or *bouncing* over the bed. They stay close together and very

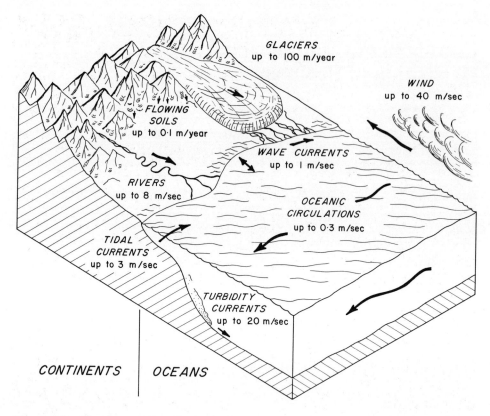

Figure 3.1 The common geological agents, where and how fast they act.

frequently collide as they are forced along by the flow. Some of the *collision forces* act upwards, against gravity, and it is they which keep the particles from being deposited. By contrast, silt and clay particles are small and light. They are transported as a *suspended load*, buoyed up by the *eddies of turbulence* in the water, like the ping-pong ball suspended on a jet of water in a shooting gallery at a Fair. The eddies push upwards, sideways and downwards, but their upwards push is the greater. The suspended sediment is distributed throughout the whole body of a river, whereas the bed load stays close to the bottom.

Commonly as much as one-half of the load transported by streams and overland flows consists of *dissolved weathering products*. These substances are obtained chiefly by the leaching of soils.

Underground waters

Underground streams occur mainly in areas of *limestone rocks*, in which they have helped to shape caves. The water is rich in carbon dioxide and entrains the limestone by slowly dissolving it away, so further enlarging the cave systems (Figure 3.2).

Figure 3.2 A tunnel shaped by an underground stream in the bedded Carboniferous Limestone of County Clare, Eire. Notice the gravel on the stream bed and the ripple-like solutional forms on the walls.

Groundwater occurs in the *pores* and *fractures* of the crustal materials. Beneath the land the groundwater layer is hundreds to thousands of metres thick, extending downwards commonly from within one metre of the surface. Below the ocean bed, however, the layer of underground water seems to be much thinner. Pores and fractures in rocks are individually small in size, but they are commonly joined together to form long pathways for the movement of underground water. This flows at a hardly noticeable velocity under the influence of pressure differences. Even clay particles are too large to be carried by groundwater, but it can transport large amounts of *dissolved substances* leached from rocks and soils. When some of these minerals are precipitated in the rock, that is, deposited chemically, the rock becomes better cemented. Many loose sediments must be changed into rocks, that is, *indurated*, partly in this way.

Waves
The waves (Figure 3.3) you have seen during sea-side holidays are mounds and hollows that travel swiftly over the surface of the water which fills the

Figure 3.3 Waves sweeping the sandy beach at Weymouth, Dorset. There has recently been a storm, when big waves swept up seaweed growing on stones from deep water.

oceans, seas and lakes. In sufficiently shallow water they give rise at the sea bed to *oscillatory currents* that flow to-and-fro over the bottom, parallel with their direction of travel. Big waves, such as those made during storms, produce bottom currents often strong enough to pick up and transport sand and even gravel. This the wave-induced bottom currents do in the same way as the flow over a river bed. Waves can entrain debris in a third way, of importance in the shaping of coastal cliffs. When a wave strikes a rocky cliff, air trapped below the crest together with the water is *forced under high pressure* into the cracks in the rock, so wedging it apart.

Tides
Tidal currents affect every part of the ocean, but are strongest in shallow seas and constricted waters. They are important agents of entrainment and transport, acting hydraulically and mechanically in the same way as river flows. In narrow channels the tidal currents are *oscillatory*, like the bottom currents due to waves, and so move debris chiefly back and forth. The tidal currents are *rotary*, however, in more open waters, the water particles travelling over oval horizontal paths back to their start. But tidal rhythms are rarely symmetrical, and there is generally a *slow drift* of water and debris in one particular direction. Tidal currents have their most obvious effects in estuaries and along the coast, but their importance also far offshore should be remembered. Thus the tide moves vast quantities of sand and even gravel over the shallow bed of the North Sea, many tens of kilometres from land.

Turbidity currents
The role of turbidity currents in the shaping of the Earth's face was only realised recently. They are major agents of sediment transport and deposition in the ocean deeps and in lakes fed by large rivers. Turbidity currents are nevertheless easy to make for oneself, though of course only on a very small

Figure 3.4 A turbidity current approximately 0.15m thick is flowing from left to right along the floor of a laboratory tank.

scale. Fill the bath at home with cold water to a depth of 10 to 20cm and let the water become still. Then pour in at the shallow end a suspension of soil stirred up in a cupful of water. You will see that the suspension, because of its *greater density*, flows quite quickly over the bottom of the bath through the clear water, as a current with a sharp head and a billowy body and tail (Figure 3.4). From this current a layer of sediment is deposited, as you will find when cleaning up! Similar but of course very much larger currents are frequently formed in oceans and lakes. They seem to begin life near the edges of the continents as huge *slumps* and *slides* (or avalanches) of soft sediment, triggered by earthquake shocks or storm waves.

Turbidity currents seem able to entrain sediment in the same way as rivers, and they appear to have cut for themselves a variety of channels and valleys on the ocean bed. Turbidity currents flow at express-train speeds and can transport in suspension gravel and sand in large amounts as well as mud. As they gradually decay, the currents spread this debris over their channels and floodplains on the ocean floor.

Oceanic circulations

The oceans are formed of numerous great mobile *water masses* differing in size, speed and flow direction. These currents are the *oceanic circulations*. They flow relatively slowly, except where squeezed against the sides of the continents or submarine mountains, but involve huge volumes of water. Locally the oceanic circulations are strong enough to pick up and transport sand or to scour a muddy sea bed. But for the most part they are only able to carry small suspended particles, for instance, wind-blown dust or the hard parts of tiny organisms that lived in the surface waters.

A more important role of the ocean currents is to transport and distribute large amounts of *dissolved substances*, the oxygen and carbon dioxide dissolved from the atmosphere and the compounds introduced by rivers draining the soil-covered land. These substances are essential to the varied forms of life in the ocean. For instance, corals and shell-fish need calcium carbonate for their hard parts, the fishes use calcium phosphate for their bones, and the tiny radiolarians need silica to make into their fragile skeletons.

Winds

The winds are currents of air demonstrating the *circulation of the atmosphere*. They are potent agents of entrainment and transport, especially where there are few plants to protect the land (Figure 3.5). Thus debris movement by the wind is especially important in the great hot deserts of low latitudes, as well as in the polar wastes.

The wind pushes and drags at particles just like a current of water and can readily pick up sand and dust. But the natural winds are rarely strong enough to entrain pebbles, which are left behind as layers protecting the underlying sediments. Much of the sand is transported by *saltation*, that is, the grains make long leaps through the air as they bounce like ping-pong balls off the particles lying still on the ground. The coarsest sand particles, however, roll

Figure 3.5   A strong wind blows up clouds of sand and dust from a newly ploughed field in East Anglia.

gently over the ground, or *creep*. They are pushed slowly forward under the repeated impact of the saltating grains. The wind transports dust in *suspension*, buoyed up by the eddying air as mud in a river. Sand grains in the wind rarely travel higher than a few metres above the ground, but the dust is often carried up into the air to heights of hundreds or thousands of metres.

Glaciers
A significant fraction of the Earth's moisture is found in the polar regions and at high altitudes in the form of *ice*, a substance with many strange properties. A piece will *shatter* into fragments as the result of a blow from a hammer; this is the kind of behaviour typical of a solid. On the other hand, a large enough mass of ice will *flow* under the action of gravity, thus behaving like an ordinary liquid. Streams of ice, or *glaciers* (Figure 3.6), are common in nature and are of several types. Glaciers travel very slowly in comparison with most other currents, generally at velocities measured in terms of metres or tens of metres per year.

Glaciers are nevertheless powerful agents of entrainment and transport. They can flow over gently sloping as well as steeply sloping surfaces and involve huge volumes of ice, their thickness measuring hundreds or thousands of metres. On account of these features, flowing glaciers apply large drag forces to their beds and so pick up particles of many different sizes. The larger pieces of rock—boulders and great blocks—are dragged off the glacier bed where the ice moves over hummocks. This debris is then used by the ice to scour or abrade the glacier bed, just as we use sandpaper to smooth wood. The products of glacial abrasion are powdered mineral particles.

Figure 3.6 A glacier flowing down a steep valley in northern Norway.

Figure 3.7 The ends of the steeply dipping limestone and shale beds in this quarry have become bent over downslope under the action of soil creep.

Flowing soils
Soil and subsoil are loose materials that form *sheet-like layers* blanketing large parts of the land. Most of these layers are in a state of *creeping motion*, travelling downslope at an average rate of a few millimetres or centimetres

each year. We call them *mass-movements* and their slow flow is readily demonstrated. Thus in quarries showing steeply dipping beds, one may commonly see that the ends of the strata have become bent over downslope (Figure 3.7). Also noticeable are *trails of stones* broken off the beds and carried downhill in the soil. Other signs of creep are trees, fences, and telegraph poles tilted from the vertical, and displaced roads.

The downhill motion of soil and rock waste occurs in several ways. In cool or wet climates the soil commonly becomes waterlogged. It may then flow slowly but continuously like a very stiff liquid. In dry hot climates, the downhill motion often occurs because of alternate *wetting and drying*, or *heating and cooling*. The soil swells on being wetted or heated, but shrinks on being cooled or dried. Hence a soil particle is repeatedly lifted up and down as the result of these changes. However, because the soil is on a slope, the particle always returns to a position a little further downslope than its starting position (Figure 3.8). The repeated *freezing and thawing* of water in the soil

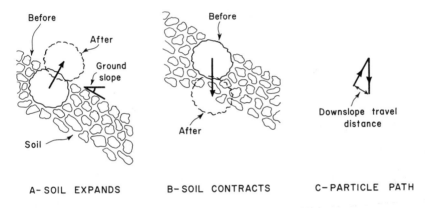

Figure 3.8 The slow downward creep of soil particles as the soil alternately contracts and expands.

acts in the same way in areas of cold climate. Each soil particle is lifted up a little when the ice crystallises beneath it. With the thaw, however, the particles return to positions a little further downslope. The growth of plant roots and the burrowing of worms may also cause downhill soil movement.

Soil movement is greatly speeded up if the base of a slope is being eroded by a river. Indeed, the downhill creeping of soils is probably the most important way in which rivers obtain sediment. At the same time the production of new soil material becomes certain, because of the steady removal by creep of the matured soil.

Rates at which the agents act
So far little has been said about the *rates* of entrainment, transport and deposition. Many tedious measurements may be required to establish these rates, but it is only when they become known that we can judge the

geological importance of the agents involved.

When entrainment occurs at some point, the Earth's surface at that point becomes lowered in elevation, because debris has been removed. Hence we measure the *rate of entrainment* by the time required for the point to be lowered in elevation by a given amount, that is, as a *velocity*. Now deposition of sediment at a place increases the elevation of that place, whence deposition may also be measured in terms of velocity. How can we tell entrainment from deposition? The answer is to use algebraic signs. During deposition the surface moved upwards, but with entrainment the movement was downwards. Calling upward movements positive and downward movements negative, we can easily distinguish entrainment from deposition but measure both in terms of velocity.

There is another way of measuring deposition and entrainment. Instead of thinking of a point on the surface, we can imagine that we have measured out a unit of area. The rate of entrainment (or deposition) is then the *weight* of debris removed from (or added to) that *area* in a given *time*.

How is the intermediate process of transport measured? When an agent carries sediment from one place to another, it does so at a definite rate with respect to a stationary observer. The rate of transport by a river or valley glacier, both of which occupy definite channels, may be measured as the *total weight* of sediment that moves past an observer on the bank in a given *unit time*. Alternatively, if we want to compare the transport rates at different places in the channel, we can measure the sediment transport as the *weight* passing along a lane of *unit width* in *unit time*.

Competence and capacity of the agents

Mention may now be made of two general properties of the agents shown in Figure 3.1. The *competence* of an agent of transport is the size or calibre of the sediment particles that agent can pick up. As you can see in Figure 3.9, the wind is generally competent only to carry sand and dust, whereas currents of water can pick up boulders as well. The other property is *capacity*, that is, the weight of sediment that an agent can carry along in each unit volume of the transporting medium, under steady conditions. Each agent has a particular capacity and in a real sense behaves as a *transporting machine*. Machines are able to do *work* only by making use of their *energy supply*.

Everyone knows the story that the idea of gravitation came to Sir Isaac Newton because one day an apple fell on his head. Rivers, turbidity currents, glaciers and mass-movements are all pulled downhill by gravity just like the apple. Hence each of these agents changes energy of position, called *potential energy*, into energy of motion, known as *kinetic energy*. Each agent as it flows is producing *power*, a part of which may be used to do the work of picking up and transporting sediment. In this repect the agent is like a lorry which burns petrol (heat energy) to make the power to carry goods along a road. As the lorry increases in power, so more goods can be carried and at a larger speed. Similarly, the capacity of a geological agent increases as its power increases. Generally the power increases with increasing velocity of

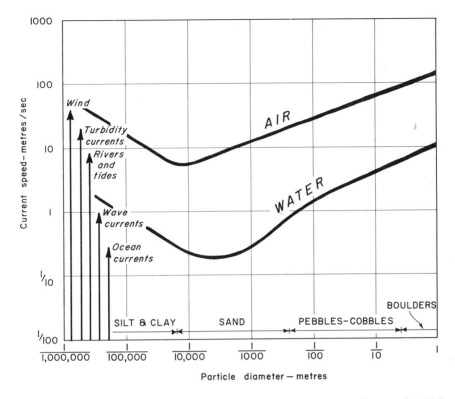

Figure 3.9 Competence of geological agents. The graph gives the speed which a current must reach before a mineral particle of a given size can be picked up. The arrows on the left show the ranges of speed for some common agents.

flow. The capacity of rivers, for instance, increases as the square of the current velocity.

The immediate source of energy of the oceanic circulations and the wind is *heat* radiated to the Earth from the Sun. Air is heated in equatorial regions, and so expands, becoming less dense. This light heated air rises, like the balloon at a Fair, and cold air flows in from the side to take its place, creating a wind. Many of the surface ocean currents are driven by the *wind*, their direction depending on the direction of the wind, the shape of the land and the Earth's rotation. Most of the deeper oceanic circulations are caused by either *heating* or *cooling* the water. In Polar regions the sea is cooled and made more dense. The dense water sinks to the ocean bed and warmer lighter water takes its place at the surface by flowing in from the side. In equatorial regions the rate of evaporation of moisture from the sea is relatively high, and here dense water is produced as the result of increasing the amount of dissolved salts. This dense water sinks, setting up further circulations.

The waves on the surface of the sea take energy from the wind. The eddies in the wind *push and pull* at the mobile water surfaces and shape it into hummocks and hollows that are driven along. Waves have some potential energy, because the water at the wave crests is further from the Earth's centre than the water at the troughs. They also have some kinetic energy, because they and the waters beneath are moving. The competence and capacity of waves increases with their height and speed.

The tide depends for its energy on the fact that the water covering the Earth is affected by the *gravitational pull* of the Moon and the Sun. The ocean surface is distorted because of the pull, in such a way that it bulges towards these heavenly bodies. But because the Earth is spinning on its axis, while the Moon orbits the Earth, the bulges appear to move round the Earth. Thus at a fixed point on the coast the level of the sea rises and falls and we have a tidal current. The capacity and competence of a tidal current increases rapidly with increasing speed.

Causes of entrainment and deposition

There is one last question to be asked. If the ability of an agent to transport sediment (capacity) depends on its power, what causes entrainment and deposition? The answer is that entrainment and deposition are caused by a suitable *change of power*. Think of a long corridor with a door at each end; imagine that you have been asked to carry two buckets of sand through the corridor, but several full buckets are already standing there. You enter one door with your two buckets, but they are heavy and your arms grow tired, so you leave one bucket in the corridor and carry out only one through the other door. You have put down (deposited) one bucket because your power or ability to do work has *decreased*. There are now more buckets in the corridor than before. Imagine that you once again enter the corridor with two buckets. This time a bar of chocolate and a cooling drink have been put for you in the corridor. After consuming them, you feel refreshed and stronger than ever, so you pick up a third bucket of sand, emerging with three buckets. This time you have entrained one bucket of sand, because your ability to do work was *increased* by the supply of energy contained in the chocolate and drink.

Natural agents of transport behave in a similar way. Where a river deepens and widens, its velocity and power decrease and sediment is deposited. As the water flows into shallower and narrower places, however, the velocity and power increase and some sediment is picked up. Similarly, when the velocity of a tidal current slackens, sediment becomes deposited. If the wind increases in speed, it picks up more particles. However, neither the wind nor a current of water can entrain sediment indefinitely, for as the particles become more concentrated the flow grows stiffer and thus slows down, depositing some of the load.

Chapter 4
The Work of Rivers and Underground Waters

Introduction

Your atlas shows that rivers great and small are widely distributed over most continents. In fact, approximately 70 per cent of the total land area is drained by rivers that reach the sea, the remainder being formed chiefly by deserts, some hot like the Sahara and others cold like Antarctica. Rivers renew the waters of the ocean. At the same time, they remove from the land the products of rock weathering, partly in solution and partly as detritus. Annually the world's rivers bring to the oceans approximately 20 billion tons of weathered matter.

Man often interferes with the geological work of rivers. Many have been dammed to provide water for irrigation, hydroelectric power, or for industrial and domestic use. Others have become seriously polluted by industrial wastes and sewage from the cities. Man has lived by rivers for thousands of years, but is still ignorant about them and their proper use.

The hydrological cycle

The water in the hydrosphere and atmosphere is called *meteoric water*, to distinguish it from *juvenile water* which reaches the Earth's surface by way of volcanic eruptions.

As the wind blows over the ocean, some of the water is *evaporated* and passes into the atmosphere as *water vapour*. The vapour can change back into liquid water, or even into ice, if the air holding it is cooled sufficiently, as often happens when the wind rises over hills or mountains. Several things

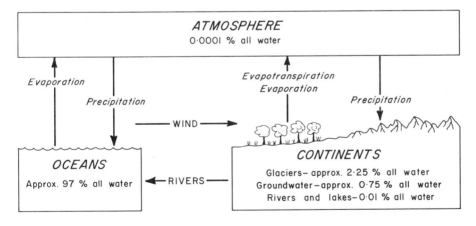

Figure 4.1 A model of the hydrological cycle, showing the movements and stores of water.

happen to this precipitate of *rain* or *snow* once it reaches the ground. Some water soaks through the soil and is added to the *groundwater*, preserved in the pores of the rocks near the surface. A second portion, flowing either through the soil or over the surface of the ground, swells the *rivers* and is returned to the oceans. A third portion may be held in *lakes* and a fourth added to *glaciers* as further layers of snow. In dry weather the land may lose water, either because of *direct evaporation* from the ground surface, or through the 'breathing' (*evapotranspiration*) of plants which absorb water from the soil.

This pattern of events is called the *hydrological cycle* (Figure 4.1). Evaporation of water from the land and ocean gives water vapour which is returned to the earth from the atmosphere in the form of rain or snow, thus completing the cycle. The cycle includes several *stores* for water, between which water particles move as time passes. The largest is the ocean itself, containing 97 per cent of all meteoric water. Ice-sheets and glaciers form the next largest store. The groundwater reservoir is a comparatively small store of water. Even smaller are lakes, rivers, and the atmosphere.

Wells and springs

In many regions a supply of water can be obtained only by digging or drilling a *well* (Figure 4.2). The well first enters the *soil*. Below is a *zone of aeration*, where the subsoil and rocks are moist but contain atmospheric air trapped in the cracks and pores. The base of this zone is formed by the *water table*, the level defined by the upper surface of the water standing in the porous rock. The water table is higher during wet cool seasons than during hot dry ones. The purpose of a well is, of course, to reach beneath the water table. Below the water table is the *groundwater zone*, the base of which ordinarily lies far below the ground surface. Dry rocks occur beneath the groundwater zone, because of the relatively high temperature found. The water table occurs at a variable

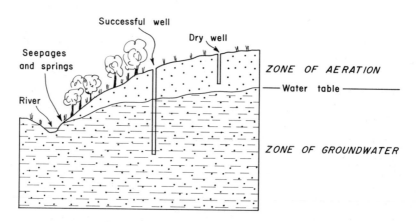

Figure 4.2 The water table standing in a porous formation, and the wells sunk to reach it.

depth below the ground surface. It lies deepest, a few metres or a few tens of metres down, where the climate is hot and dry and the rocks allow the free passage of water through them. At *seepages* and *springs*, however, and along the edges of lakes and rivers, the water table crops out at the surface of the ground.

Wells are best sunk into rock formations that are *porous* and *permeable*. Formations of these kinds can hold large amounts of water—they are known as *aquifers*—and allow an easy flow of groundwater. A porous formation may have substantial amounts of space between the particles of which it is composed, as in a layer of sand or gravel. Alternatively, there may be a large number of interconnected open fissures in the rock, as is the case with the Chalk. A permeable rock is one in which the fissures or pore spaces are comparatively large and interconnected, so as to form long passage-ways for groundwater. Sand and gravel and the Chalk are also good examples of permeable deposits. Beds of shale, on the other hand, illustrate *non-porous* and *impermeable* formations called *aquicludes*.

Permeable and impermeable rocks are often combined together in *geological structures* that favour the outcropping of the water table at particular places, in the form of seepages and springs (Figure 4.3). Each impermeable

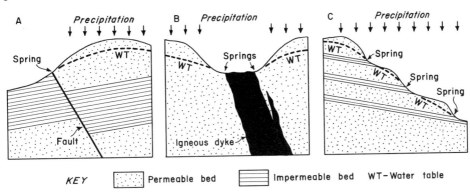

Figure 4.3 The occurrence of seepages and springs at the outcrop of the water table is closely controlled by geological structure.

layer tends to block the downward movement of groundwater, and so diverts the flow sideways with the result that water is forced to the surface. Lines of springs and seepages commonly for this reason mark the positions of faults, igneous dykes and sills, and the outcrop junctions of permeable and impermeable beds.

A very important structure is formed when an aquifer between two impermeable horizons becomes shaped by earth movements into a basin-like fold (Figure 4.4). The rain that falls on the outcrop of such an aquifer flows down-dip towards the trough of the fold, where the groundwater is under *hydrostatic pressure*. This pressure increases in amount as the outcrop of the aquifer grows in elevation above the trough and the aquifer itself fills up. The

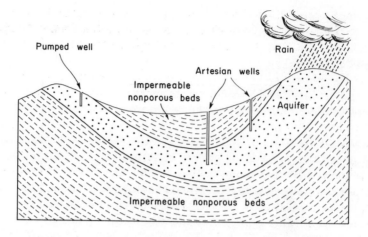

Figure 4.4 An artesian basin in vertical section.

pressure may be so great as to cause water to spout from wells penetrating the aquifer in the basin centre. A well spouting in this way is known as an *artesian well* and the basin-like fold in which artesian wells are sited is called an *artesian basin*. The Chalk which passes from the North Downs beneath London to the Chiltern Hills is an aquifer in an artesian basin, though so much water has been drawn from it that wells in London must nowadays be pumped. Artesian basins occur in many other parts of the world, for instance, beneath the deserts of Saudi Arabia and Australia.

Limestone caverns
Limestone is an unusual kind of rock because it is relatively soluble in water. It is not surprising that the slow flow of groundwater through cracks and joints in limestone beds forms *cave systems* below ground and, on the surface, leads to distinctive land forms known as *karst*. Periods of time measuring hundreds of thousands or even millions of years are required by these processes, which seem to work as follows (Figure 4.5).

Some of the rain falling on the ground sinks down towards the groundwater reservoir. In filtering through the aerated zone, it dissolves carbon dioxide gas from the entrapped air, and so becomes weakly acid. This acidified water mixes into the uppermost layers of the groundwater reservoir, which are in continuous but very slow motion—a rate of a few metres per year is typical. As the acidified water flows along, it *chemically dissolves* calcium carbonate from the sides of the fissures in the limestone, so gradually opening them out into large interconnected passages and chambers. These form a cave system, though they are not yet free to the air. The caves lie within a narrow zone in the limestone, because the acidified water is restricted to the uppermost part of the groundwater reservoir. Deeper in the reservoir the water is saturated with calcium carbonate, and so cannot make caves by solution. If for some reason the water table is lowered (e.g. change of climate), the caves are abandoned and cave formation begins again at a lower horizon. Repeated shifts of level of the water table can create complicated systems of galleries, chambers and shafts ranging vertically through many hundreds of metres of the limestone rock.

A lowering of the water table allows plants and animals to penetrate the caves. In these caves, abandoned by the groundwater reservoir, one finds calcium carbonate being *precipitated* rather than dissolved. These precipitates take strange and beautiful shapes. From the roof hang long pendants called *stalactites*, formed over long periods by the precipitation of calcium carbonate from drips of water. Rising from the floor below to meet the stalactites are thicker columns called *stalagmites*. A *pillar* occurs where stalactite and stalagmite have fused together. Less often calcium carbonate is precipitated in the form of folded *curtains* or in rippled *sheets* on the rock.

As cave systems grow larger, some of their shafts are extended upwards to meet water draining from the ground. Each shaft as it enlarges begins to attract more of the surface drainage, and around its mouth a large funnel-shaped hollow, called a *sink-hole*, begins to form. Sink-holes are very common

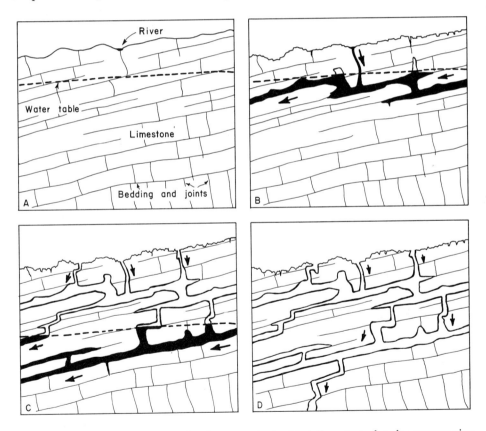

Figure 4.5 A model of cave development in bedded limestone by the progressive descent of the water table.

in limestone districts. A good example is Gaping Ghyll on the south-east slopes of Ingleborough in North Yorkshire. Another kind of sink-hole arises by the collapse of the roof of a large chamber thinned and weakened by solution. Limestone districts with plentiful sink-holes, and streams running underground in caves rather than over the surface, are said to show *karst topography*. A wide range of karst features exist in Britain where the Carboniferous Limestone outcrops.

River drainage basin and net
Turning now to rivers that flow over the land, you will see from your atlas that each river takes water and sediment from a definite area, known as its *drainage basin*, separated from adjacent similar areas by hills or mountains forming *watersheds* (Figure 4.6). You will also see that the river channel

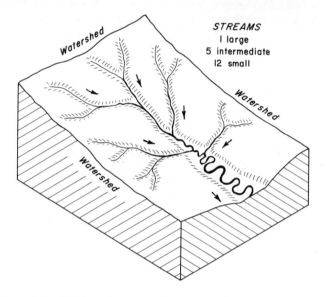

Figure 4.6 Model of a river drainage basin and drainage net.

branches so as to form a characteristic pattern, the *drainage net*. Typically, the different parts or *segments* of the channel are arranged like the twigs, boughs and trunk of a tree. In the drainage net of a large river there is a large number of small tributaries, a smaller number of segments of intermediate size, but of course only one trunk stream. The small tributaries are generally also the shortest and their beds slope more steeply than the channel segments of intermediate size, which in turn are steeper than the trunk stream. The length and slope of each channel segment is therefore appropriate to the relative position of the segment in the drainage net.

In practice a variety of patterns can be found amongst drainage nets, just as the pattern of twigs and branches differs between one kind of tree and another. In some drainage nets the channel segments tend to join at right angles, because the river was developed on rocks that were folded, faulted and tilted. The folds and faults, along which the rock has been weakened, have guided the directions assumed by the different segments. Other nets show a large angle between segments, as in the oak tree, whereas some show a small angle, a feature of the branches of the poplar.

In order to understand rivers we should know about the amounts of water passed by the channels at different places in the drainage net. This amount, known as the *discharge*, is measured at special places on the channel, called *gauging stations*, usually in units of cubic metres per second. To make the measurement at a gauging station, soundings are taken to find the cross-sectional area of the flowing water, and then the average velocity of flow is estimated using a current meter. The discharge is the cross-sectional area (in square metres) multiplied by the flow velocity (in metres per second). The discharge can be studied in two complementary ways.

Firstly, the discharge can be measured daily throughout the year at a particular station on the river. Thus we obtain the *hydrograph*, a curve showing how discharge varies with time. Rivers in hot dry climates often have a short flood season and a long period of low flows, on account of seasonal rains. A short flood season usually marks rivers also in cold climates, due to rapid melting of the winter snows. Rivers in humid and temperate regions, however, do not show such marked discharge variations.

Secondly, we can measure the discharge on one particular day but at many gauging stations. Such a study would show that the smallest tributaries had the smallest discharges and the main stream the most water.

There is one value of the discharge at each gauging station which cannot be exceeded without the river spilling over its banks to flood the surrounding land. This value, called the *bankfull discharge*, is important in planning measures to prevent floods.

Channel shape and pattern
Segments of river channels when viewed in plan are found to be either *meandering* or *braided* in pattern.

Braided rivers (Figure 4.7) are found usually to be very broad and shallow, with steeper and wider beds than meandering ones of similar discharge. In a braided stream the water is divided between a large number of flat-bottomed channels which branch and rejoin around lozenge-shaped masses of sediment, called *braid bars*, deposited by the flow. The channels and bars shift rapidly as the currents scour in some places but deposit in others.

Meandering rivers (Figure 4.8) are more common than braided ones. The solitary channel swings from side to side in regular loops ordinarily called *meanders*. Each loop encircles a mass of sediment, called a *point bar*, laid down by the river. The channel in each meander loop is roughly triangular in cross-section, the concave outer bank rising steeply above a deep *pool*, as the inner convex bank slopes up gently to the surface of the point bar. Between loops, where the channel alters direction of curvature, the river cross-section is broad, shallow and flat-bottomed. This point is a *crossing*.

Figure 4.7 Gravel bars in the braided channel of the River Naver, Sutherland, Scotland.

Figure 4.8 An oblique aerial view of the alluvial valley of the meandering River Jordan 40km north of the Dead Sea, Jordan. Note the ox-bow lakes on the floodplain.

The regularity of the channel loops is a striking feature of meandering rivers, and may be measured as the *meander wavelength*, the down-valley distance between consecutive loops. Studies made on natural rivers and on streams in the laboratory show that the meander wavelength bears a definite relationship to size of stream. The wavelength increases in proportion to the square root of bankfull discharge.

Stream sedimentation: channel and floodplain
Rivers deposit sediment partly in their *channels* and partly on the adjacent flat lands called *floodplains*. Generally, the *channel deposits* are sands and gravels, because these materials typify the river bed-load, transported close to the channel floor. The *floodplain deposits*, however, are chiefly clays, silts and the finest sands. Only these materials are fine enough to be carried in suspension, and so spread far from the channel during floods.

Deposition in meandering streams takes place on the convex bank of each meander loop, in harmony with erosion of the steep concave bank (Figure 4.9). Hence the meander grows larger and cuts its way sideways through the floodplain deposits. This movement is proved by the sediment ridges, or *scrolls*, which decorate the tops of point bars. Each scroll records a major act of deposition by the stream and an earlier position of the channel. The river behaves like this because the current in each bend follows a *corkscrew pattern*. The flow at the channel bed is inwards, from the pool to the shallows, and so pushes sediment on to the point bar. A complementary outward flow occurs

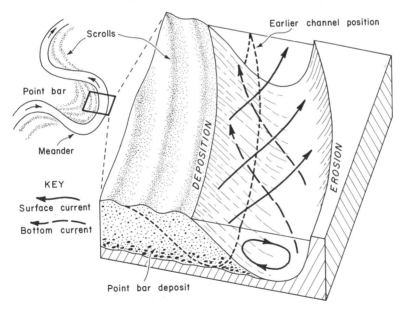

Figure 4.9 Model of a meandering stream to illustrate the erosional sideways movement of the meanders and the corkscrew pattern of water flow around each loop (compare the beaker experiment).

just below the surface of the water, to make the spiral complete. You can easily model this process by very gently stirring a large beaker of water using a steady circular movement. A few crystals of potassium permanganate dropped in the water will bring out the spiral pattern very clearly. If sand is

Figure 4.10 A violent flood in the Finke River system of central Australia left these sand dunes (with superimposed current ripples) on the channel bed. When dug into, the dunes proved to be cross-bedded inside. They are approximately 3m in wavelength.

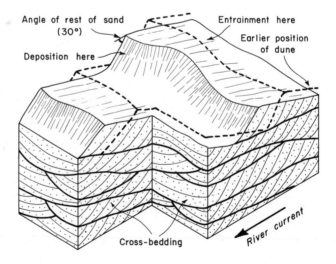

Figure 4.11 Model of dunes on a river bed, their movement and internal cross-bedding.

put in the beaker, and the water stirred more vigorously, you will find that the spiral current pushes the grains towards the centre of the bottom of the beaker. The pile of grains is a miniature point bar.

Sometimes two meander loops approach so close that the river during a flood cuts across the narrow neck of land between them, so taking a new and shorter course. The *cut-off channel* loop becomes plugged with sediment at the ends and forms an *ox-bow lake* on the floodplain (Figure 4.8). More sediment is introduced with each flood and so the lake gradually changes into a peaty *swamp*.

Natural point bars comprise sand and gravel, the coarsest material lying at the base of the bar above an erosion surface swept out by the migrating river. Some of the sand may show *parallel laminations*, formed when the bed was smooth and flat. Cross-bedded sand may occur elsewhere in the bar. This kind of deposit forms when large *dunes* on the river bed move downstream under current action (Figures 4.10, 4.11). They are series of large sand ridges with crests at a steep angle to the current. In a large river their height may be several metres and their wavelength a hundred metres or so. The river washes sand from the gentle upstream side of each dune and deposits it on the steep downstream slope, where the grains *avalanche* to form inclined layers which may become fossilised as *cross-bedding. Current ripples* often occur on river beds. They travel downstream in the same way as dunes and are dunes in miniature, with a height of a few centimetres. Internally the ripples show *cross-lamination*, a small scale form of cross-bedding.

River floodplain deposits reflect a combination of *aqueous* and *atmospheric processes*. As river discharge increases, the water first spills over low gently sloping ridges bordering the channel. These ridges of sediment, called *levees*,

Figure 4.12 Polygonal cracks due to the drying up of mud in a shallow pool. Birds paddled in the soft mud before the pool finally dried out.

are slightly eroded as the floodwaters race down them towards lower lying areas. As the floods deepen, however, the suspended fine particles begin to sediment, the levees and floodplain receiving a new layer of mud and sand. The thickest layers are deposited on the levees, which are thus made taller and steeper. With flood recession these deposits become affected by atmospheric processes. Wind and sun dry the mud, causing it to shrink and crack, forming *mud cracks* (Figure 4.12). Plants recolonise the floodplains, sending their roots into the sediment; and animals once more roam the surface, leaving footprints in the plastic mud. In dry climates *wind-blown sand dunes* are found on river floodplains. Under hot conditions, the salts carried in solution in the floodwaters become precipitated in salt pans and on clay flats.

River terraces and drowned valleys
Often the relative level of land and sea remains constant a long time. The rivers are then found to have greatly widened their valleys and to have partly infilled

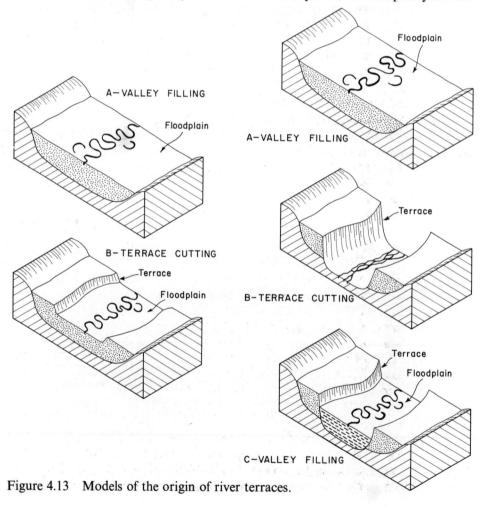

Figure 4.13   Models of the origin of river terraces.

them, constructing broad floodplains. But the rivers will abandon these floodplains and build new ones at lower levels in the alluvial fills if sea-level—the base-level for the rivers—falls in relation to the land for some reason. In this way an abandoned floodplain becomes a *river terrace*, a landform consisting of an *escarpment* or *bluff* roughly parallel with the river and a *platform* or *flat*, formerly the floodplain. This sequence of events and the resulting landforms appears on the left-hand side of Figure 4.13, the final picture being of a river floodplain flanked by a terrace perched on the valley sides.

The same morphology results if the river base-level moves down and then up relatively to the land. This sequence appears on the right-hand side of Figure 4.13. A downward movement of base-level is accompanied by terrace-cutting, whereas valley-filling goes hand-in-hand with an upward movement. Clearly the two situations shown in Figure 4.13 can be told apart only if the *stratigraphy* of the valley fill is known in detail.

However, *fluctuation of climate* over a river drainage basin is an equally common cause of terrace formation. A change of climate alters the amount of water discharged through the drainage net and affects the modes and rates of weathering and soil-formation over the basin, which in turn change the amount and coarseness of the debris carried away. These factors interact in a complex way, but their final effect is often to alter the shape of the longitudinal profile of the trunk stream and its larger tributaries. If the longitudinal profile is made to become more strongly concave to the sky, the river will entrench itself in a shallow valley, or cut a terrace within an existing but incomplete valley-fill. Studies made in many parts of the world show that entrenchment (or *dissection*) is most likely to happen when the climate changes to *drier*, so that the amounts of water and sediment discharged by the streams become reduced. When the longitudinal profile is made flatter, however, the river will fill up with sediment, or aggrade, the valley it occupies. Older terraces bordering the river may become buried in the process. *Aggradation* is the usual result of the climate changing to more *humid*. Such a change increases stream discharge. Because of the greater availability of water, it also enhances weathering and soil-formation over the basin, and increases the amount and coarseness of the sediment carried by the streams.

River terraces are found all over the world. All large rivers in Britain display a number of terraces, for example, the Clyde, Trent, Severn and Thames. In London the Thames has three terraces. The highest, called the *Boyne Hill Terrace*, lies about 80 metres above sea-level. The next lowest is known as the *Taplow Terrace*, and it is about 45m in elevation. The lowest terrace of all, the *Floodplain Terrace*, is 15 to 20m above sea-level and borders the modern floodplain. Terraces are important economically because they preserve river sands and gravels useful for building and road-making.

When sea-level stood exceptionally low, rivers cut valleys extending far out from present shores. These valleys are now *drowned* and partly infilled with marine sediments. Their floors lie deep below sea-level. Thus the former course of the River Elbe in north-west Germany can be traced far out across the bed of the North Sea. In the Thames Estuary there lies a partly buried gorge which the river occupied when sea-level was much lower than today.

There are several reasons why the relative level of land and sea may change. During ice ages, sea-level falls because a great deal of water becomes locked up as glacier ice. When a thick ice-sheet melts, a load is removed from the crust, which consequently becomes uplifted. At the same time sea-level rises because of the reappearance of the water. The relative level of land and sea may also change on account of tectonic movements within the Earth's lithosphere.

Deltas and coastal plains

Much of the sediment carried by the world's large rivers goes to form vast *deltas* and *coastal plains* constructed close to base-level. In Britain, however, recently formed deposits of these kinds are quite unimportant.

A *delta* resembles in plan the Greek letter of that name. Each is formed by a single river which has been divided into smaller channels, called *distributaries*, through which the water and sediment is dispersed. Your atlas will show that large deltas are widely distributed, from the humid tropics to the frozen shores of the Arctic Ocean. Many deltas have dimensions measured in tens or hundreds of kilometres.

Deltas vary chiefly in shoreline character (Figure 4.14). The Niger delta (West Africa) illustrates a type marked by a smoothly curved sandy shore

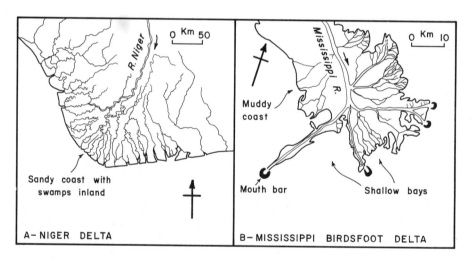

Figure 4.14 Mississippi birdsfoot and Niger deltas compared.

associated with tidal estuaries, swamps and lagoons. This type forms only where rivers empty sand into a sea affected by waves powerful enough to carry the grains easily along the shore, away from the river mouths. The Mississippi delta (Gulf of Mexico) illustrates another type, known as the *birdsfoot* after its shape. Each distributary now dumps most of its load on a

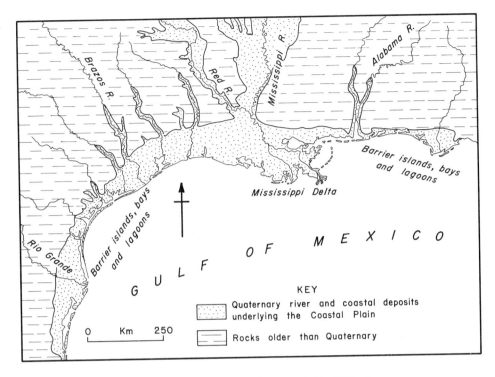

Figure 4.15 The coastal plain of the north-west Gulf of Mexico.

submerged bar at its mouth, because waves and currents are too weak to carry sediment far. Hence the distributaries rapidly advance seaward on finger-like bodies of sediment. The areas between fingers remain largely unfilled and contain shallow marine bays and brackish swamps.

River-borne sediment is often deposited close to the edge of the sea in the form of *coastal plains of alluviation*, which are larger even than deltas. These plains are constructed by several rivers acting together. Good examples occur on the east coast of the USA, between the Hudson River and Florida, and on the north-east coast of Sumatra in the East Indies. The coastal plain encircling the Gulf of Mexico is one of the largest in the world (Figure 4.15). It includes a complex of deltas formed at various times by the Mississippi River, and has a shoreline in places sandy with bays and lagoons and in others marshy and muddy.

Chapter 5
The Work of the Wind

Introduction

The wind does most geological work in the regions of land called *deserts*, some of which have *cold* and others *hot* climates. It also plays an important part in building up *coastal sand accumulations*, which share many features with the regions commonly regarded as deserts.

A desert is hard to define precisely. Unfortunately, temperature is not a reliable pointer to a desert, nor is the scarcity of plants and animals. Most scientists agree, however, that a region is a desert if it is *large in area* and has a relatively *low rainfall*, the mean annual value being 25cm or less. The deserts of greatest geological interest have high temperatures, at least during daylight hours, sparse vegetation and very few animals. This chapter is mainly about deserts of this kind, of which the Sahara in North Africa is an excellent example. However, the Earth has other deserts with a very cold climate, the great ice-sheet of Antarctica being one. These are best discussed when explaining the geological work of ice. Comparatively few of the coastal areas of wind-action are true deserts, even though the wind is the dominant agent and the vegetation sparse.

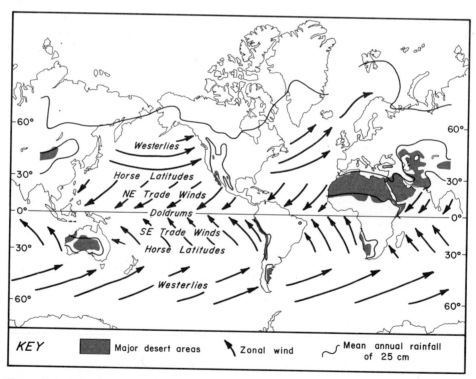

Figure 5.1 The distribution of hot deserts in relation to rainfall and the zonal winds.

Occurrence of deserts

Deserts depend on climate and their occurrence reflects the distribution over the Earth of the major *climatic zones* (Figure 5.1). The distribution of these zones in turn is controlled by the way the Sun's heat energy becomes distributed through the atmosphere.

The air above the equator is heated to a higher temperature than the air near the poles, where the Sun's rays have merely a glancing path to the ground. The heated air is less dense than its surroundings, and so rises. This creates a region of low pressure, into which flows cooler air from neighbouring regions of higher pressure. In this way the atmosphere is set in circulation on a very large scale, giving what are called the *zonal winds* (Figure 5.1). In the equatorial regions we find a belt of low-pressure air known as the *Doldrums*. The *Trade Winds* blow towards it from zones of high pressure known as the *Horse Latitudes*, lying about 30° latitude north and south of the Equator. Poleward of the Horse Latitudes lie the *Westerlies* which blow towards belts of low-pressure occurring about 60° latitude north and south of the Equator.

The dry air that descends at each of the regions of high pressure can precipitate little moisture. Hence the largest deserts lie between 10° to 30° latitude north and south of the Equator (hot deserts) and in Polar latitudes (cold deserts). The deserts of Australia and Saudi Arabia, like the Sahara, are excellent examples of hot deserts. Good examples of cold deserts, in addition to Antarctica, are provided by the Greenland ice-sheet, Alaska, and Siberia. The central parts of Asia are deserts mainly because they lie at a very great distance from the sea, the source of the moisture transported by the wind. They are marked by an extreme continentality. The Patagonian Desert in South America is close to the sea but lies in the *rain-shadow* of the

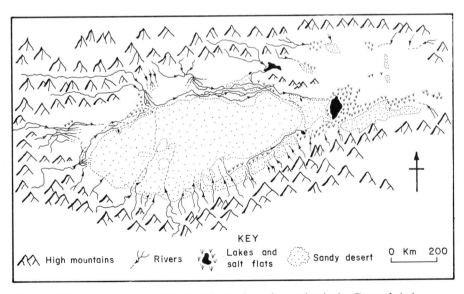

Figure 5.2 Sketch map of the Takla Makan desert basin in Central Asia.

Westerlies blowing across the Andes from the Pacific Ocean.

Stream drainage in deserts

Although hot deserts receive little rain on average, evidence of *stream drainage* can usually be found within their boundaries. Commonly the drainage system is *internal*, the valleys and channels directing the water towards the interior of the desert rather than to the sea beyond. For instance, internal drainage is shown by the Takla Makan Desert in Central Asia and by the Lake Eyre Basin in Australia (Figure 5.2). The drainage channels are dry for most of the time, because of the scarce rainfall. However, water frequently lies at shallow depths beneath their beds.

Rain seldom falls in deserts. When it falls heavily in one area, perhaps a few times a year or every few years, *violent floods* arise which race down the channels and valleys as steep walls of water. On reaching flatter ground these occasional floods spread out rapidly as *sheet flows* from which gravel, sand and mud are thickly deposited. In the desert hills and mountains, fresh surfaces of rock are exposed as the result of scouring action of the floods, and so weathering processes become speeded up for a short period.

Figure 5.3 An oblique aerial view of rugged hills surrounded by pediments in the Jordanian Desert.

Mountains, hills and pediments

The character of the upland parts of deserts depends on the structure and nature of the underlying rocks and on the climate. Hills and mountains formed on dipping rocks are jagged in appearance, with sharp peaks and steep sharply ridged sides. Flat-lying beds in deserts form table-like hills, sometimes called *mesas*, with cliff-like gullied sides.

Usually each hill or mountain mass is surrounded by a broad, smooth area called a *pediment* sloping gently away in all directions (Figure 5.3). Within the hills and mountains occur steep-sided, flat-bottomed valleys called *wadis* which channel the floodwaters (Figure 5.4). Most wadis empty

Figure 5.4 A desert wadi emptying on to an alluvial fan, Trucial Coast.

on to extensive *fans* of gravel, sand and sometimes mud built by the floods across the surface of the pediment. Often shallow channels pointing to stream action can be found on these fans. Because gravel is the material chiefly found on these fans, we have here a special kind of desert, the *stony desert* (Figure 5.5). Many parts of the pediment, however, lack widespread deposits of loose sediment and are formed from bare rock. Such areas form another kind of desert surface, the *rocky desert*.

Figure 5.5   Stony desert in the Trucial Coast.

Figure 5.6   A collection of ventifacts from various modern deserts. The pebbles are a little less than natural size.

Stony deserts are generally underlain by a single layer of closely packed stones that covers beds of sand and gravel and sometimes mud. The stone-covered surface may be due to the action of the wind, which has winnowed away the sand and mud brought in by the floodwater, but left the larger gravel, or to sorting by running water. A deposit of stones left behind in this way is often called a *lag*, and the process whereby the wind selectively removes the sand and dust is called *deflation*. This process can easily be studied by visiting a sand beach or a recently ploughed field in a sandy area when the wind is strong.

Strangely shaped stones called *ventifacts* characterise some stony deserts. Ventifacts show a number of relatively flat smooth and often wrinkled faces that join along sharp edges (Figure 5.6). These faces were cut by the abrasive action (sand-blasting) of the particles blown along close to the ground by the wind. The fact that more than one face can be found on each ventifact suggests either changes of wind direction, repeated overturning of the stone (perhaps because of flooding), or wind-eddying in the lee.

The areas of rocky desert include wide level stretches of *bare rock* on which may stand isolated *pillars* and *towers* of rock. Often the joints in the rock, measuring many metres apart, have been widened and deepened by wind action and weathering, so that the desert surface has a large-scale tessellated appearance. Locally in many deserts, and particularly around the Tibesti Mountains in the Sahara, the wind has shaped the outcropping rock into a variety of large features which lie parallel to the wind. Those called *yardangs* are elongated ridges of rock measuring tens of metres in length which are tallest and steepest at the up-wind end. In places the sides of the ridges are undercut, showing that sand-blasting is most effective close to the ground, where the flying grains are most concentrated. Much larger than yardangs, and locally covering huge areas, are groups of long parallel *ridges and furrows* carved in the rock. These ridges measure hundreds or even thousands of metres in length and are spaced tens or hundreds of metres apart. Such ridges and furrows curve gently around the Tibesti Mountains, where they are a common feature, and thus show how the wind is deflected by this obstruction on the desert floor.

Sand seas
Sand seas form a comparatively small part of most deserts. These seas occur where the wind over long periods has heaped together large amounts of sand obtained through deflation elsewhere in the desert. They reveal a wide variety of *sand formations* built up by the wind in transporting the sand, by *saltation* and by *creep*. These formations are *ripples*, *drifts*, and *dunes*. The saltating grains take long flights through the air and travel at a speed similar to the wind itself. On striking the ground at the end of one flight, a saltating grain may either bounce back up into the air or splash up a number of fresh grains. The creeping grains, however, are slowly rolled along under the impact of the saltating particles. Because of the powerful impacts, the particles forming desert sand seas are very well rounded and usually frosted.

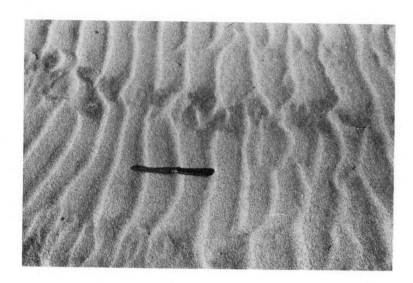

Figure 5.7 Wind ripples in sand, afterwards lightly pitted by rain-drops. Table-knife gives scale.

Figure 5.8 In the Culbin sand hills, Elgin, Scotland. Wind-blown sand has become trapped as drifts to the lee of tufts of grass.

Ripples are the smallest of the formations of loose sand shaped by the wind (Figure 5.7). They occur in all deserts and can also be found amongst the sand hills of the British coast. Each rippled surface shows a series of parallel ridges of sand lying at right-angles to the wind. They usually measure about a few millimetres in height and 5 to 10cm between crests. The wind drives the ripples along by eroding grains from the gentle upwind faces and depositing them on the steeper downwind sides. A sufficiently strong wind will smooth out the ripples, driving sand over a level bed.

Drifts are larger than ripples but generally smaller than dunes (Figure 5.8). Usually drifts form as long ridges of sand trailing downwind of rocky outcrops, isolated boulders or stones, and tussocks of vegetation. Some of the largest drifts can be found in the lee of rocky cliffs. Locally drifts have been built against cliffs that face the prevailing wind. These are known as *climbing drifts*, because grains can travel over the drift up to the level of the cliff-top.

The *barkhan dune* has the most perfect shape of any formation of blown sand

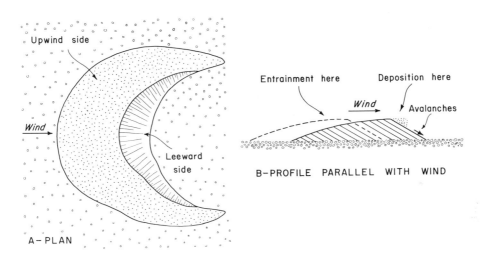

Figure 5.9 Desert barkhan dune and its movement beneath the wind.

(Figure 5.9). It is a large crescent-shaped mound measuring tens or hundreds of metres across the base and with a height commonly of many metres. The upwind side has a gentle slope. The leeward side lies between two large ridges of sand which project downwind from the dune. It slopes down steeply (about 30° from the horizontal), at the *angle of rest* of the sand. Each dune travels with the wind in a similar way to the dunes on river beds. Sand is entrained from the upwind side and deposited on the sheltered leeward face, down which the accumulated grains periodically flow as *avalanches*. Thus the sand inside a dune shows the structure we have called cross-bedding. Each inclined layer in this structure dips in the same direction as the prevailing wind and represents an avalanche on an earlier leeward slope.

Dunes of other kinds may be found where the sand is thickly spread over the desert surface. *Transverse dunes* (Figure 5.10) resemble wind ripples in shape and relationship to the wind but are enormously larger. They are commonly many metres in height and hundreds of metres in wavelength. Cross-bedding is found inside them. *Longitudinal dunes*, called *seifs* in the Sahara, are often very large (Figure 5.11). They are long parallel ridges of sand stretching for tens or hundreds of kilometres. Their spacing apart varies from a few hundred metres to several kilometres, and the largest are

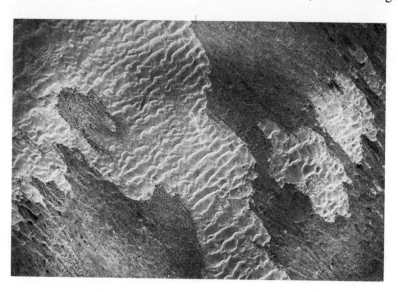

Figure 5.10 A vertical aerial view of part of the desert of Iraq. Several large spreads of sand carry transverse dunes of wavelengths measuring tens of metres.

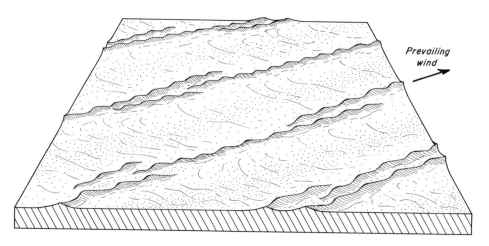

Figure 5.11 A model showing longitudinal dunes in a perspective aerial view.

truly sand mountains, reaching heights of hundreds of metres. Seif dunes lie *parallel* to the direction of the prevailing wind, and in this respect differ from barkhan and transverse dunes, the lee slopes of which are aligned at steep angles to the air stream. The explanation for the orientation of seif dunes is thought to be that, in the air flowing over the heated desert surface, there are very large *corkscrew-like eddies* arranged parallel to the wind and rotating in

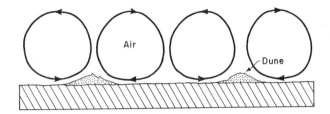

Figure 5.12 A vertical transverse section through the atmosphere above a field of longitudinal dunes, to show the possible wind pattern responsible for the dunes. In this cross-section, which is two-dimensional, the air particles travelling in the corkscrew vortices appear to be moving in circles, carrying sand over the desert floor from the flat areas between the dunes up on to the dune flanks. Using wire bent into open coils, and Plasticine or Plaster of Paris, try to make a three-dimensional model of the dunes and the suggested wind flow.

opposite directions (Figure 5.12). The seifs appear to form where the flows in adjacent eddies converge over the ground, so heaping up the sand into a long ridge. Where the flows in adjacent eddies diverge over the ground, the sand is removed and so we can obtain the *corridors* between the dunes.

The last kind of dune deserving mention is the *star-shaped dune* (Figure 5.13). These are regularly spaced mounds of sand rising often to a height of 500m above their surroundings. Their steep sides are marked by sinuous sharp-crested ridges that meet at the dune summit. These ridges thus appear

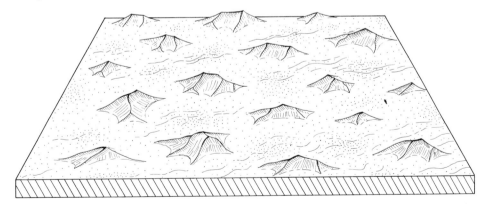

Figure 5.13 A model showing a field of star-shaped dunes in a perspective aerial view.

61

from the air like the rays of a star, and so give this type of dune its name. Little is known about the conditions necessary for the formation of star-shaped dunes, but possibly they form where the wind blows equally strongly from nearly every compass point. Several sand seas in the Sahara are dominated by these enormous dunes.

Coastal belts of wind-blown sand
Coastal accumulations of wind-blown sand occur in all parts of the world, except for the humid tropics where the rank growth of vegetation limits wind-action. The sand is obtained by the deflation of the adjacent beaches and travels inland with winds blowing off the sea. It is easy to tell when a beach is being deflated. Firstly, you may actually be able to see the sand blowing over the beach surface in the form of *wavering streamers*. Secondly, you may find places where the bedding within the beach sand has been delicately etched out by the scouring wind. Thirdly, if deflation has taken place, each pebble or shell on the beach will lie at the upwind end of a small sand ridge which it has protected from erosion.

The sand scoured from the beach is blown inland over short distances to build up *dune ridges*. Usually the dunes are irregular in shape, due to the action of the relatively numerous plants growing in the sand. These plants act both to *trap* sand being carried in the wind and *bind* the grains already deposited. Although lacking the regularity of form so marked in the desert, the dunes of the coast commonly reach many metres in height.

Coastal sand-dune belts occur in many parts of Britain, for example, along the shores of the Moray Firth (Scotland), on Merseyside between Liverpool and Southport, at several places on the Welsh coast, and in north Norfolk.

Chapter 6

The Work of the Sea: Cliffed and Sandy Coasts

Introduction
The modern coastline shown in atlases is important to geologists as well as to geographers. It separates land from sea, dividing the face of the Earth into one part exposed to atmospheric processes and another washed by the hydrosphere. If we can learn how to tell coastal features preserved in rocks, maps can be drawn showing past arrangements of land and sea. These are called *palaeogeographical maps*, because they tell about the geography of the past, and you will no doubt have studied examples.

But the coastline is not a fixed boundary. It shifts in position with every change of sea-level, often to the extent of many kilometres where the land is low-lying and flat. The coastline will also change position if debris is added to or removed from it, due to the action of the sea, a river, or the wind. Generally speaking, rocky coastlines shaped into cliffs change more slowly than low-lying coasts underlain by sand or gravel, because rocks are relatively difficult to erode.

Agents affecting coasts
These are waves, tidal currents, and the wind. They may either add or remove debris from the coast, according to local conditions. Rivers may also affect coastlines, but generally in an indirect way.

The *wind* has most effect on low-lying sandy coasts in areas of warm dry climate, particularly when it blows strongly for long periods off the sea. Sand scoured from the beach is blown inland to form rows of tall dunes. Regular dunes aligned at right angles to the wind occur where the climate is so dry that few plants grow in the blown sand. In moister areas irregular dunes appear, because there are more plants to trap and bind grains. The plants grow so thickly and so close to the sea on humid tropical coasts that the wind has practically no effect. The wind by dragging on the water may produce weak currents in the sea near the shore.

Tidal currents also affect coasts. The *tide* is the rhythmical up-and-down motion of the surface of the sea (Figure 6.1). It represents the movement round

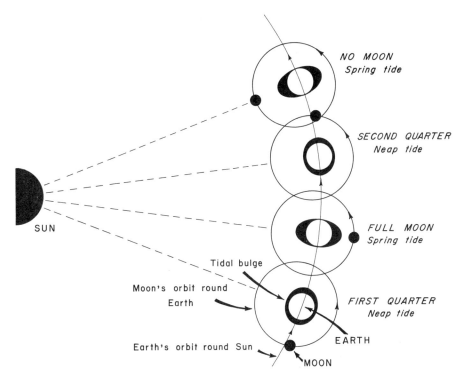

Figure 6.1 The tide is generated as the Moon rotates around the Earth (once every twenty-eight days), while the Earth spins on its axis (once each day), and both orbit the Sun (once each year).

the Earth of a bulge on the surface of the ocean, caused by the *gravitational pull* of the Sun and Moon on the mobile water. Along with the vertical motion of the sea surface there are horizontal movements of the water—the tidal currents—which are strongest in shallow or restricted parts of the sea. High tide generally occurs twice each day, but in some areas, like the Gulf of Mexico or the coast of Korea, is observed only once daily. The levels reached by low and high tides are not constant, but change with the phases of the Moon, the difference between levels being the *tidal range*. The range is greatest (*spring tides*) when the Sun and Moon are in line, at new moon and at full moon. The range is least (*neap tides*) when Moon and Sun act at right-angles. Spring tides occur twice each lunar month, and in British waters generally have a range of 5 to 10m.

Everyone during a seaside holiday has noticed *waves* travelling on the surface of the sea (Figure 6.2A). Each wave has a definite *height* measured as

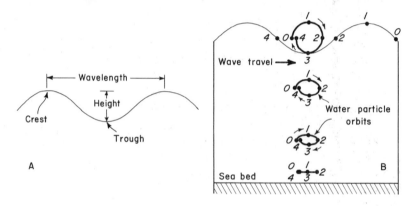

Figure 6.2 Water waves. A—the wave in profile can be described in terms of height and wavelength. B—the motion of water particles beneath travelling waves. As the wave form travels from left to right a water particle occupies successively the positions 0, 1, 2, 3, 4 on the water particle orbits.

the vertical distance between *crest* and *trough*, together with a *wavelength*, the horizontal distance from one crest to the next. Waves also have a definite *speed*. In deep water this grows with wavelength. When a wave enters shallow water, however, it slows down as the water depth grows less, because of friction between the moving water and the sea bed.

You will probably have seen that waves of several different sizes exist together on the surface of the sea. The big waves have smaller waves on their backs, and on these small ones there are still smaller waves. Hence we can say that a *spectrum* of waves is present. Recalling that waves are made by the combined pushing, pulling and shearing action of the wind on the mobile sea, how is this observation to be explained? The answer is that the larger waves, to which the name *swell* is often appropriate, are likely to have travelled from storm centres hundreds or thousands of kilometres away, where they were made by violent winds. The local wind conditions are in no way reflected by their size, for one can often observe swell on the calmest of days. However, the waves of intermediate and small size you observed are likely to have been formed by the locally acting wind. You should find that they have an orientation consistent with the local wind. Waves made by the local wind may nevertheless approach swell in size if the wind is very strong and persistent.

The motion of waves over the sea represents a movement of water in the same direction as the waves. If waves meet a coast, the cliffs or beach will act like a dam, blocking the movement of water with the waves. Hence at a coastline the waves *break* and the water moving with them is turned sideways and/or backwards, giving rise to various *coastal currents* (Figure 6.3). *Longshore currents* moving parallel to the shore form where waves approach

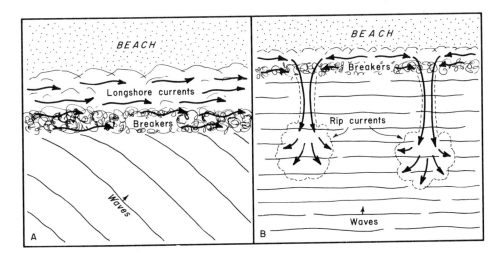

Figure 6.3 The origin of coastal currents. In A the waves approach the beach obliquely, giving powerful longshore currents. In B the waves approach steeply, setting up longshore currents and intermittent rip currents.

obliquely. Where waves approach steeply, however, these currents are relatively weak and water tends to pile up against the beach or cliffs. The piled-up water is periodically released as narrow *rip currents*, which flow for tens or hundreds of metres out to sea as dangerous undertows. These two kinds of wave-generated current are important in the movement of sediment on and near beaches.

Cliffed coasts
Cliffed coasts appear where the edge of the land consists of resistant materials, *cliffs* being steep rock-faces generated by destructive storm waves. As each wave strikes the cliff, a huge mass of water and entrapped air is flung at the

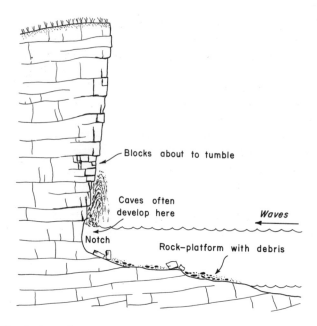

Figure 6.4 Vertical section through a sea cliff shaped by wave-action.

Figure 6.5 An oblique aerial view, looking south, of cliffs cut in the London Clay, Herne Bay, Kent.

lower part of the cliff, giving a violent blow. Large pressures are exerted in the cracks of the rock, pieces of which become dislodged. In addition waves fling quantities of sand and shingle at cliffs and the rocks thus become smoothed and scoured.

Cliff profiles reflect these processes (Figure 6.4). At the foot is a broad flat wave-scoured *rock-platform*, partly exposed at low tide and often thinly covered with loose debris. A shallow *notch* appears at the base of the cliff. Wave-action is most effective here, and as the cliff base is cut back, the rocks above may *collapse*. Blocks tumble a few at a time from undercut cliffs of well-jointed rocks. Huge *landslips* only slowly worn away by the sea occur along cliffs formed from soft and poorly jointed beds, especially shales and glacial clays.

The general shape of a cliffed coast depends on the relative resistance of the rocks in the cliffs, the structure of the beds, and the general intensity of wave-action.

Cliffs of relatively poorly resistant beds (Figure 6.5) generally are straight or gently curved in overall plan and shallowly embayed. Commonly they show *landslip scars* and aprons of debris (often with *mudflows*) representing partly worn landslips. Good examples appear in the London Clay along the Thames Estuary, and in East Anglia where there are thick glacial clays.

Where rocks are hard and well-jointed but flat-lying, the cliffs present many long *headlands* and deep *embayments* (Figure 6.6). Towers of rock,

Figure 6.6 Embayed cliffs and a sea-stack formed of Old Red Sandstone, Yescanaby, Orkney.

called *sea-stacks*, may lie in front of such cliffs, and often the headlands are pierced by *caves*, the roofs of which may in time collapse to give sea-stacks. The Old Man of Hoy, in the Orkney Islands, is a well known sea-stack no less than 130m tall. The Old Red Sandstone in north-east Scotland forms cliffs of this general type.

Along some coasts the rocks are alternately strongly and weakly resistant. The resistant layers then form *promontories*, the weaker beds being cut back as *bays*. These effects are well shown by the cliffs of folded rocks of the St David's Peninsula, Dyfed (Figure 6.7). The resistant rocks here are igneous,

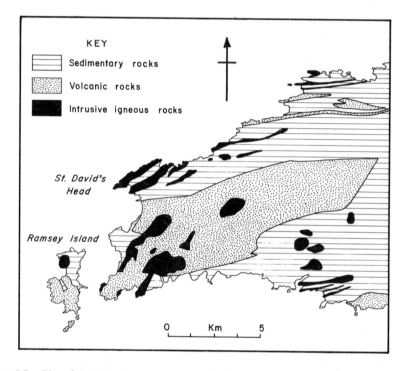

Figure 6.7 Simplified geological map of St David's Peninsula, South Wales. The form of the coast depends on the geological structure and on the occurrence of weak and resistant beds.

and they mainly form the headlands. The bays between are sculptured chiefly in the weaker slates and shales. Notice how the shape of the whole Peninsula is controlled by the structural grain.

In places a steeply dipping layer of resistant rock forms a *wall* protecting softer beds to landward from wave attack. One such area lies on the Dorset coast, where long stretches of the cliffs are formed by the resistant Portland Beds of Jurassic age (Figure 6.8). In several places these Beds have been pierced and worn sideways by the sea. At Stair Hole waves forced a tunnel

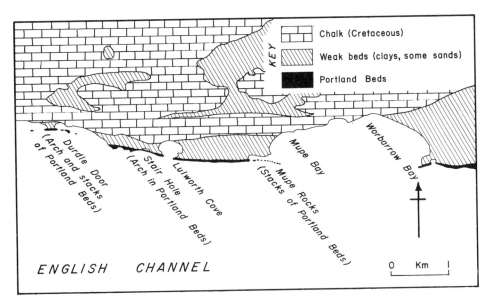

Figure 6.8 Simplified geology of the Dorset coast east of Weymouth. Differential marine erosion of steeply dipping weak and resistant beds.

through the Portland Beds and hollowed out some of the weaker rocks behind. Lulworth Cove seems to represent a more advanced stage of erosion, for any arch cut early into the Portland Beds long ago collapsed and was widened out. The weak Cretaceous clays were rapidly cut back, the sea now beating on Chalk cliffs. A further stage is represented by Mupe Bay and Warbarrow Bay, where long stretches of the wall of Portland Beds have been demolished.

Sandy coasts
In many regions the coast is low-lying and formed of loose and easily transported debris. This is often the case where rivers bring sand and gravel to the sea, as on the shores of *deltas* and *coastal plains of alluviation*. Locally sand deposited along the coast has been supplied by the scouring of the sea bed well off shore.

A low-lying sandy coast borders the eastern shores of the North Sea, between the Chalk cliffs of Boulonnais and Denmark. This coast is affected by strong tidal currents and powerful waves, the sediment drifting northwards and eastwards along the shore. The sand and mud deposited there in recent times has come partly from the rivers, for example, the Rhine and Elbe, and partly by the scouring of drowned glacial deposits. The most obvious feature is a string of large islands built up by sea action (Figure 6.9A). These serve as a *barrier* sheltering the relatively calm Wadden Sea to their east and south from the storm-tossed North Sea to the north and west.

The islands consist largely of *sand* washed up on the wide beaches facing the North Sea. It is clean and well sorted and is heaped up into many different features indicative of current action. On parts of the beach the sand is deposited over smooth flat areas, in a series of flat even *laminae* dipping seaward, each just a few grains thick. Such a deposit would make a flaggy sandstone if cemented and fossilised. Elsewhere the beach shows ripple marks or is shaped into low *sand bars*. The wind often blows strongly onshore,

Figure 6.9 Tidal environments compared. A—barrier islands and intertidal flats of a part of the North Sea coast of the Netherlands. B—intertidal flats and barrier islands of a part of the Niger delta, West Africa.

carrying sand inland to form tall dunes on the islands.

Between each pair of islands lies a narrow channel, as much as 50m deep, allowing the tide to ebb and flood between the North Sea and the Wadden Sea. The mouth of each channel on the seaward side is encircled by large submerged sand banks crossed by channels, forming what is called a *tidal delta*. The large channels branch into smaller ones within the Wadden Sea, just as a tree divides into branches or a river into tributaries. Strong tidal currents sweep these channels, the beds of which consist of sand and gravel accompanied by transported shells representing bivalve molluscs and gastropods living in the Wadden Sea.

The Wadden Sea is quite shallow. In fact, well over half its total area is uncovered at many low tides, the uncovered part forming a broad *intertidal flat*. These flats are formed chiefly of sticky grey mud inhabited by a relatively few worms, molluscs and crustaceans. Tidal channels meander through the mud flats, reworking them in a way similar to rivers meandering through alluvium. Just above high-tide level on the shores of the Wadden Sea one finds level *salt marshes*. These are meadows inhabited by salt tolerating plants, since the marshes are drowned by the higher tides and during storms. Scattered on the surface of the marsh are shallow depressions, called *salt pans*, in which sea water may lie for a time.

The Wadden Sea and its barrier islands are worth comparing with the coastal barrier that borders the Niger delta, West Africa (Figure 6.9B). This barrier also comprises a chain of sandy islands built up by the sea and serving to protect a large intertidal flat from the ravages of the ocean. The differences depend on climate. The Niger delta lies in the humid tropics and dense jungle covers the barrier islands. Hence there are no wind-blown dunes on the Niger delta barriers, even though the beaches are sandy and strong winds prevail. For the same reason the intertidal flat bears a luxuriant growth of *mangrove* bushes and trees able to grow in the salty mud. These plants, with wigwam-like systems of aerial roots, stabilise the intertidal mud flats and make them resistant to erosion by tidal currents. Consequently the tidal channels crossing the flats are connected like twine in a woven fishing net. In the Wadden Sea, where the flats lack vegetation, the tidal channels are tree-like.

Beaches, spits and tombolos
Beaches represent sediment piled against the shore by wave and tidal action. They are of course found along the sandy coasts just described, but may also be seen along cliffs and in bays within cliffed coasts.

Several distinct parts can usually be recognised in the vertical profile of the beach and adjacent areas (Figure 6.10A). Offshore there may be *longshore bars* on which waves periodically break. The *foreshore* lies between tide marks, being repeatedly covered and uncovered by waves. Foreshore slope is largely controlled by the coarseness of the beach sediment and by wave strength. Thus gravel beaches slope on average at about 15°, whereas sand beaches rarely dip more steeply than 5°. Provided the beach sediment is constant, slope increases with increasing intensity of wave-action. The highest point

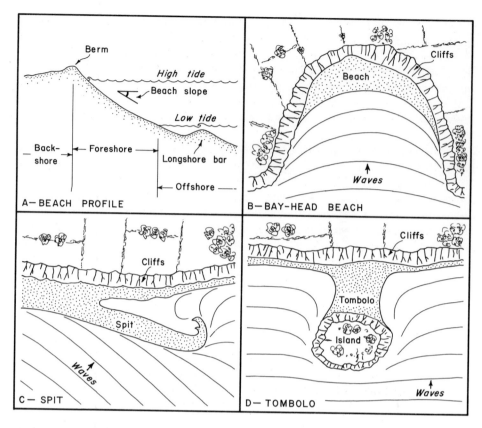

Figure 6.10 Models of coastal features shaped by wave-action.

on the beach profile is known as the *berm crest*. This represents the greatest height to which storm waves will fling debris. The *backshore* lies to landward of the berm crest. It may be either level or slope gently inland. Only storms and the wind supply sediment to the backshore.

Sediment is carried along beaches in two ways, by *longshore currents* and by wave *swash and backwash*. The longshore currents partly underlie the line of breaking oblique waves—the *breaker zone*—where the water is violently agitated and sediment is constantly stirred up from the sea bed. The transport of particles by wave swash and backwash depends on what happens after waves break. Some of the water from the broken wave surges forward up the beach as the *swash* and down again as another surge called the *backwash*. When waves reach the beach obliquely, the swash moves obliquely up and across the beach, so carrying debris in steps along the shore in the same direction as the adjacent longshore currents.

Because waves are slowed by friction as they enter shallow water, a wave approaching the beach at an angle is moving slower inshore where the water is shallow than offshore where it is deep, the crest therefore bending towards the shore (Figure 6.3). This bending of waves as they pass into shallow water is called *wave refraction*. It helps to determine the shape and positioning of beaches and other large coastal accumulations.

Many cliffed coasts are shaped into bays at the heads of which there commonly are smoothly curved *bay-head beaches* (Figure 6.10B). From the air the curve of such a beach matches very closely the curve of the crests of waves in the bay. How can the beach have been formed? The point to notice is that each wave entering the bay is refracted in the shallows near the headlands. When there was no beach, the waves arrived at an angle to the head of the bay, and so carried sediment along the shore from the direction of the headlands. As more debris became available, the beach grew in size and began to conform more and more closely in shape to the curve of the refracted waves. When the beach and the refracted waves conformed exactly in curvature, longshore sediment transport stopped, because the waves no longer arrived obliquely. Such a beach cannot change further, except in size, and so has acquired an *equilibrium shape*.

The longshore transport of sediment by waves may produce at favourable places a *spit*, a long bank of sand or shingle attached to the coast at one end only (Figure 6.10C). Several conditions favour production of spits.

The spit known as Orford Ness, at the mouth of the River Ore in Suffolk, illustrates one of these conditions (Figure 6.11A). This is an example of a spit developing at the mouth of a river or tidal inlet which empties on to a coast where there is a vigorous longshore transport of debris. The spit results from a battle, so to speak, between the water striving to escape from the channel and the sediment obliged to move along the coast past the channel mouth. Consequently the mouth is deflected in the direction of the longshore transport,

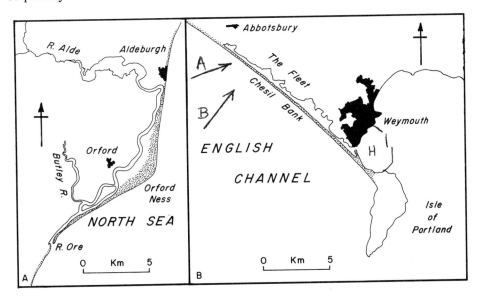

Figure 6.11 Coastal features produced by wave-action, A—Orford Ness, and B—Chesil Bank.

73

as you can see in the case of the River Ore.

An interesting gravel spit lies off Blakeney, Norfolk. This ends today in a large complex hook, constructed by waves refracted around the end of the spit. For many centuries sediment has been drifting westwards along the coast, causing the spit slowly to move bodily in that direction. Hence to landward one can see hook-shaped bodies of gravel each of which represents the end of the spit in an earlier position further east.

Some spits grow across the mouth of a bay or broad estuary, leaving open to the sea only a narrow channel. Good examples occur at the mouth of the River Exe in Devon, at the entrance to the Barmouth Estuary in Wales, and at the mouth of the Esk in Scotland.

One special kind of spit should be mentioned. This is the *tombolo*, a sand or gravel bar joining the mainland to a nearby rocky island around which waves are refracted (Figure 6.10D). A good example, formed of sand, can be seen joining St Catherine's Isle to the mainland at Tenby in South Wales. The Chesil Bank on the Dorset coast is another fine tombolo, in this case very large in size and built up entirely of gravel (Figure 6.11B). Chesil Bank may have begun life as a spit which did not become a tombolo until it grew so far to the west that it joined the mainland near Abbotsbury.

Chapter 7

The Work of the Sea: Continental Shelves

Introduction
Today the continental masses are bordered by a comparatively narrow terrace called the *continental shelf*, submerged beneath a shallow sea. The shelves are relatively flat smooth areas stretching out from the coast as far as the *shelf-edge*, where water depth is 100 to 200m and the bottom deepens fairly abruptly down towards the ocean depths. Continental shelves are important to man because of the fisheries they support and the problems of ship navigation and petroleum exploration they raise. Hence scientists began at an early date to study the shelves.

Learning about continental shelves
The easiest way to learn about the shape of the sea bed is to map out water depth using sea-level as a datum. The older methods of mapping involved using a long graduated wooden *pole* (very shallow water only) or a *lead-line*, a heavy weight attached to a strong cord or wire lowered over the ship's side. Often tallow was smeared on the weight, to secure a sample of bottom sediment. The lead-line is slow and unreliable, however, and nowadays *echo-sounding* gives us accurate and continuous profiles of the sea bed.

The *echo-sounder* is an instrument consisting of a *source* of sound and a *receiver* combined with a chart-recorder (Figure 7.1A). The source, attached to the ship's bottom, sends down vertically a narrow beam of sound pulses, which travel through the water at a speed of about 1500m per second. The

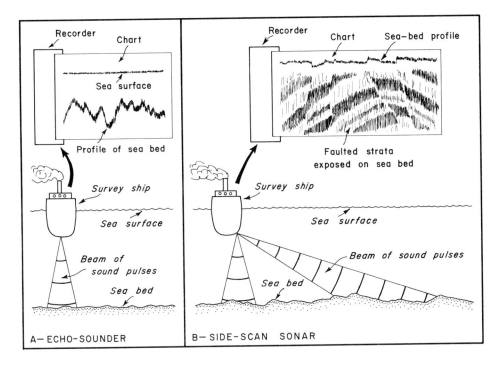

Figure 7.1 Exploring the shape and character of the sea bed using, A—the echo-sounder, and B—side-scan sonar.

pulses are reflected from the sea bed and returned to the ship, where they are detected by the receiver. It measures the time between the transmission and detection of each pulse, and on the chart records the result as a line representing the profile of the sea bed below the ship's track. If a sea-floor map is wanted, many echo-sounder profiles must be carefully fitted together. This is rarely easy because of doubts about the positioning of the ship.

One partial solution of the problems of mapping is the *side-scan sonar*, related to the echo-sounder (Figure 7.1B). This apparatus directs some pulses vertically down from the ship and others outwards at a shallow angle. These oblique pulses are therefore reflected from a broad strip of the sea bed commonly many hundred metres wide. The chart in the recorder now shows the profile obtained using the vertical beam of pulses, combined with a *map* of the strip of sea floor which reflected the inclined beam. The map is in black and white and shades of grey. The dark areas represent parts of the sea bed that reflect sound well, for instance, sand or gravel patches or rocky cliffs facing the ship. A muddy bottom, or a part of the sea floor sloping away from the ship, reflects little sound and shows up pale grey or white.

Shelf sediments are easily sampled. A *dredge* made of a steel box or steel netting may be towed behind a ship to recover rocks and stones. Large samples of sand and mud may be collected using a *grab*. One simple kind consists of two metal cups that close together like a set of teeth on touching the sea bed. However, the most useful tool of all is the *corer*, a hollow pipe fitted with guiding fins that is dropped on to the sea bed at the end of a cable. The corer pierces the bottom and traps within a cylindrical core of sediment, which may afterwards be removed. Cores several metres long are easily obtained. The core is a long undisturbed vertical slice through the bottom sediment that reveals something of the geological history of the sea bed.

Currents in shelf seas are generally stronger than in the deeper ocean. Surface currents may be estimated by tracking *floats*, but those lower down are best measured using a recording *current meter*. This instrument has a torpedo-shaped body and fins at one end to keep it facing the current. Mounted on the body is a small propeller turned by the current. The speed of the current is measured by the rate of turning of the propeller, and the direction by the orientation of the instrument. Using several current meters it is possible to find out how the currents change in space and time.

Diving is an important means of study. A diver using a *breathing tube* (schnorkel) can easily work in water a few metres deep, studying and photographing the bottom and collecting sediment samples. Furnished with *SCUBA* (self-contained underwater breathing apparatus), he can work as deep as 50m.

General character of continental shelves

The continental shelves are very varied. The largest shelf, dotted with islands great and small, lies off the Russian shores of the Arctic Ocean and is many hundreds of kilometres wide and very flat. Shelves of intermediate breadth (about 200km) occur off Argentina and Uruguay, around the Gulf of Mexico, and off Nova Scotia and Newfoundland. Portions of many shelves are partly land-locked, for example, the shelf around the British Isles, the Yellow Sea bordering China, and the Arafura Sea between Australia and New Guinea. By contrast, other shelves are no more than a few tens of kilometres wide. Shelves as narrow as this occur off the west coast of North and South America, where deep-sea trenches lie close to land. The shelf off the east coast of Africa is also narrow, but lacks a bordering trench.

Agents affecting continental shelves

At the present day these are (a) waves, (b) tides, (c) mainly wind-driven oceanic circulations, and (d) glaciers and drifting ice-bergs. They are responsible for transporting and depositing the shelf sediments and locally may erode the sea bed.

Various *wave-induced currents* appear when waves travel over the sea (Figure 6.2B). Imagine we can mark a water particle on the surface and follow its movement as waves pass by. The particle is highest when at the wave crest. As the wave travels on, the particle moves forward and downwards. Backward and downward movement begins when the wave has travelled onward one-quarter of a wavelength. The particle is lowest when lying in the wave trough, but after this has passed, it begins to ascend, moving first backward and finally forward. Notice how the particle moves over a *very nearly circular orbit*, lying a little further on from its starting position after the wave has passed.

What happens beneath the surface? Provided the water is neither too deep nor too shallow compared with the distance between wave crests, the particles will be found to follow very nearly perfect elliptical orbits becoming flatter and shorter as depth increases (Figure 6.2B). Clearly the effect of waves is to cause the water to move slowly but steadily in the direction of their travel, a current known as the *wave-drift*. Particles at the bed travel simply to-and-fro (*oscillatory motion*) on a path at right-angles to the wave crests, without up and down movement. For a given water depth, the strength of the bottom currents increases as the wave height and spacing grow larger. These currents are often strong enough to disturb sand and even gravel on the sea bed, and hence may build *wave-current ripples*. These ripples are long sharp-crested ridges of debris with a symmetrical cross-section (Figure 7.2).

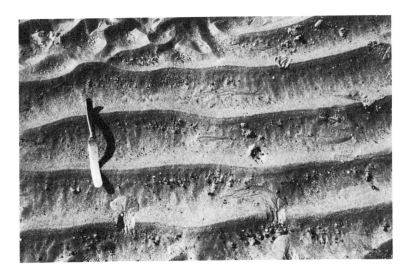

Figure 7.2 Wave-current ripples left by the tide on a sandy beach. Table-knife gives scale.

As the current flows to-and-fro over the bed, the debris moves first in one direction and then in the other across the ripple crests, thus accounting for the symmetrical form.

Where the water is very deep relative to wave spacing, the oscillatory bottom currents are too weak to disturb the sediment. This happens when the water depth roughly equals one-half the wavelength. The depth in question is known as *wave-base*, and it is clearly different for each set of waves.

Hence in calm weather, when waves are small, only the shallowest parts of the shelves are affected by wave-action (Figure 7.3A). But when there is a storm, and large waves appear, the sea bed several tens of metres down may be swept by strong wave currents, because wave-base has moved deeper

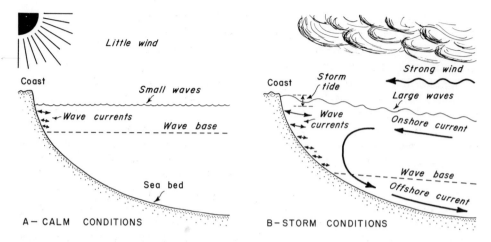

Figure 7.3 A model of wave-action on the continental shelf, illustrating the contrast between, A—calm conditions (small waves, weak currents), and B—storm conditions (large waves, strong deep currents).

(Figure 7.3B). The onshore wave-drift now becomes important, but is turned into an offshore current at the coast, which acts as a barrier. If the storm wind at the same time blows towards the land, water becomes piled against the coast, creating a *storm tide* and an additional offshore current. Thus by combining the stirring action of storm waves with offshore currents, large amounts of sand and mud may be carried away from the shallows into deeper water.

Continental shelves are also affected by *tidal currents*. Because the shelves are more open, these are less powerful than tidal flows in estuaries and inlets, and are generally weakest in the deepest water. Also, the tidal currents of open seas slowly change direction as time passes, on account of the spinning of the Earth, so that in one tidal period the flow-direction changes through one revolution on the compass-card.

Glaciers and *drifting ice-bergs* generally affect only those continental shelves found in high latitudes. Where the water is too shallow for the ice to float freely, the glaciers carve valleys and trenches into the sea bed that correspond to the valleys cut by glaciers on the land. The ice-bergs on melting drop stones and sand on to the shelf, and they may groove the sea bed where run aground by the wind or currents.

Continental shelves since the last glaciation
In the last few million years (Quaternary times) the Earth has experienced rapid oscillations of climate between warm and cold. The continental shelves as seen today owe many features to these changes.

During the cold or *glacial periods* some of the Earth's moisture was stored up in polar glaciers. Hence sea-level fell all over the world as the glaciers grew in size. As the climate warmed up, during *interglacial periods*, much of the ice melted and the sea correspondingly rose. We are at present living at a time of mild climate, resembling an interglacial period. Today sea-level is 100 to 150m higher than it was some 15 000 years ago during the last glacial episode. This range of sea level seems typical of the changes between earlier glacial and interglacial periods.

These sea-level oscillations naturally caused the shoreline to move to-and-fro across the continental shelves. When sea-level was low, during glacial episodes, the shoreline lay near the shelf-edges, and the rivers crossed the shelves in deep valleys they had cut. The shelves became weathered and soils appeared. The rivers had mouths close to the shelf-edges, and hence dumped sediment directly on to the steep sides of the ocean basins. The picture was different during interglacial periods, such as we may live in today. The shelves and their river valleys lay drowned beneath a shallow sea, the coast forming tens or hundreds of kilometres back from the shelf-edge. Thus a *marine transgression* (or drowning of the land) took place as the climate warmed up and sea-level rose. But as each glacial period was approached, a *marine regression* (or drying out of the land) took place with the fall of sea-level.

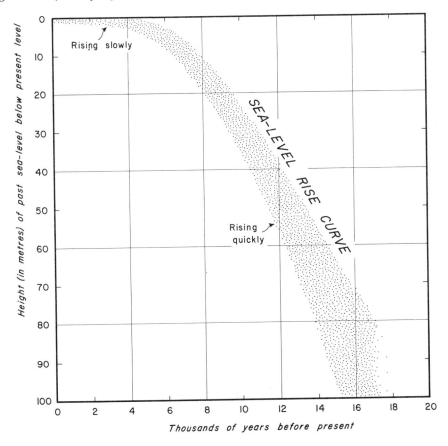

Figure 7.4 Rise of sea level since the last glaciation, based on the radiocarbon dating of shells and wood.

The most important marine transgression followed the last glacial period. It is known in some detail because geologists have been able to measure, using *radioactive carbon* (a kind of atomic clock), the ages of animals and plants that lived close to sea-level at different times during its rise (Figure 7.4). The graph shows that 15 000 years ago sea-level stood about 100m lower than today. It appears to have stopped rising about 3000 years ago.

Continental shelf in the north-west Gulf of Mexico
This is a good example of a broad continental shelf affected by mainly weak currents (Figure 7.5).

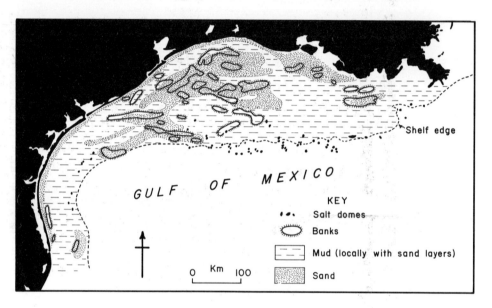

Figure 7.5 The continental shelf in the north-west Gulf of Mexico. Distribution of features of sea-bed topography and types of sediment.

The shelf is a smooth *sediment-covered plain* showing low *banks* parallel with the modern coast. The taller banks are the eroded tops of *salt domes*, huge intrusions of rock-salt which rose upwards from a great depth through the strata because of buoyancy. The less tall banks seem to be *barrier beaches*, formed in the earlier stages of the transgression and then drowned as the sea rose higher. Also there are broad shallow *valleys* and *channels*, which may represent former rivers or estuaries.

In the northern part of the Gulf of Mexico, the wind blows mainly from the south or south-east, setting up currents towards the shore and westwards along it. The waves are generally small and with little effect on the bottom away from the zone of breakers. The seasonal *hurricanes* produce large waves, however, which stir up the bottom at considerable depths. Storm tides 2 to 3m above ordinary sea-level are encountered at the coast, and large parts of the low-lying land become drowned. Much sediment travels out to sea in the storm currents.

The shelf sediments are *sand* and *mud*. The sand forms a layer a few to many metres thick lying like a seaward-dipping blanket over the whole shelf. This layer accumulated in successive strips at and near the shore as the sea rose, and it rests on red to brown soils formed during the preceding glacial episode. The sands locally fill up old river valleys or estuaries, the shoreline features on the upper surface of the layer having already been mentioned. The muds lie on top of the sands and are therefore younger. They were brought in by the rivers, chiefly the Mississippi in the east and the Rio Grande in the west. Although locally many metres in thickness, they have buried the older sand layer in only a few places. Within the muds are thin beds of sand which could represent storms.

Nigerian continental shelf
This shelf (Figure 7.6) borders the equatorial Atlantic Ocean and differs in

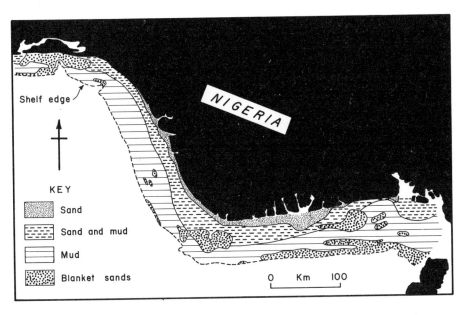

Figure 7.6 The Nigerian continental shelf, West Africa. Distribution of types of sediment.

several ways from the previous one. Firstly, it is only one-third to one-half as wide as the shelf in the Gulf of Mexico. Secondly, the Nigerian shelf has the stronger wave and tidal currents. The waves come in from the south-west and set up vigorous longshore currents. An ocean current—the Guinea Current—flows gently from west to east along the shelf. Finally, the River Niger introduces large amounts of both sand and mud.

Four kinds of sediment occur on the shelf. The oldest sediments are *blanket sands* deposited during the post-glacial transgression. Drowned coastal features can be found on the upper surface. Most of the shelf is occupied by muds and sands deposited in the late stages of the transgression and afterwards when sea-level was stable. These sediments become finer grained with increasing water depth and decreasing strength of the currents. Clean *sand* occurs on the beaches and near the shore in water depths up to 10m. From this depth down to about 50m, the bottom is underlain by alternate layers of *sand and mud*. The sand layers grow thinner when traced seaward, and could represent sediment washed from the coast during tropical storms. At depths exceeding about 50m, where currents are very sluggish, the bottom sediment is a thick layer of sticky *mud*. Locally 40 to 50m of mud and sand caps the blanket of transgressive sand.

Hence on this shelf we find that the contemporary sediments are coarsest close to shore, where currents are strongest, becoming progressively finer grained further out where currents are less powerful and the depths are greater. Many shelves do not show this horizontal sequence, because the transgressive sediments remain largely unburied.

Continental shelf surrounding the British Isles
This partly land-locked shelf (Figure 7.7) illustrates sedimentation controlled by tidal currents, with frequent assistance from waves. It is relatively shallow, large areas having a water depth less than 50m. The strong tidal flows play their part all the year round, with the fastest currents exceeding one metre per second in narrow waters and near the coast. The large waves shaped by the frequent storms make currents that can stir up sand, shells and even small pebbles in depths as great as 50 to 100m. These currents help the tidal flows to entrain and transport debris.

The shelf sediments have various origins. Most were scoured up by the sea from deposits laid down on the shelf by rivers and glaciers during glacial times. Some debris was brought by rivers after the last glaciation. The erosion of coastal cliffs and rock outcrops on the sea bed gave small quantities of sediment. Another small amount consists of the hard parts of shell-fish and other animals that lived in the sea now drowning the shelf.

The character of the sea bed depends on the strength of the tidal currents, and it is possible to trace a definite sequence of kinds of sea bed from places where the currents are strongest to areas where they are weakest. Also, sediment is being eroded in the places of strongest currents and transported to the areas of weaker currents, much of it being deposited on the way.

The areas of strongest current are being *scoured*, so that the underlying *bedrock formations* are laid bare (Figure 7.8). Erosion picks out the hard bands of rock as ribs on the sea bed, but forms hollows where the rocks are soft. In places there are large deep trenches lying parallel with the direction of the strongest tidal currents. The only loose sediment found in these areas is *gravel* and *broken shells* in small patches and in hollows between the rocks.

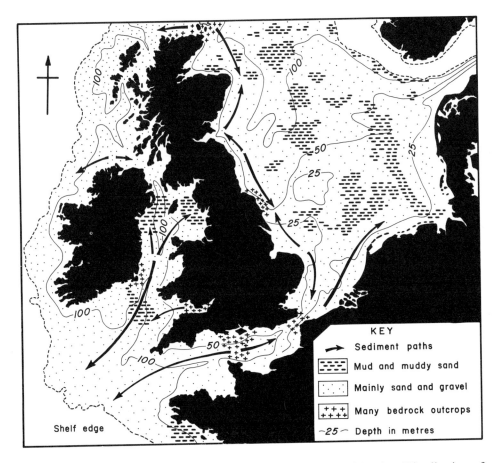

Figure 7.7 The tide-swept continental shelf around the British Isles. Distribution of types of sea bottom (greatly simplified) and paths of sediment bed-load movement.

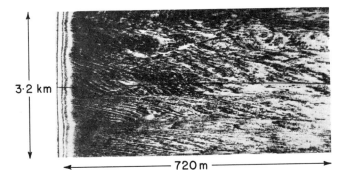

Figure 7.8 Side-scan sonar record of faulted and folded steeply dipping strata exposed on the sea bed, approaches to Cork Harbour, Eire.

Where the currents are a little less strong, the sea bed is underlain by sheets of *shelly gravel*. Moving over the flat top of the gravel is a small amount of *sand*. The side-scan sonar shows the sand arranged in numerous large *ribbons* parallel with the direction of the strongest currents (Figure 7.9).

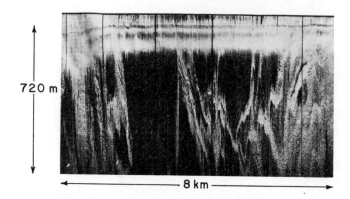

Figure 7.9 Side-scan sonar record of rather irregular sand ribbons on a gravelly sea bed. Near Bardsey Island, Cardigan Bay.

These ribbons are only a few centimetres thick, but they can be up to 15km long and 200m wide.

In the next zone down the path of the sediment the currents are moderate in strength. Here the sea bed is underlain by a thick layer of *shelly sand*. The sea floor is no longer flat and smooth but shaped into huge sand banks and a variety of large whaleback hills, known as *sand waves* and *dunes*, whose crests lie at steep angles to the strongest tidal currents. These features travel with the currents and are easily picked out on the chart made by an echo-sounder or side-scan sonar (Figure 7.10). The smallest have a spacing of 10m and are

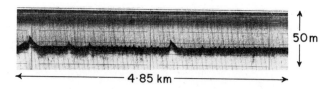

Figure 7.10 Echo-sounder record of sand waves on a flat sea floor, Cardigan Bay. The water is up to 35m deep and the tallest sand waves are nearly 10m high.

no more than one metre high. The largest measure 25m in height and 1000m in wavelength. No wonder the first scientist to discover these sand waves compared them with desert dunes! Cross-bedding seems to be preserved inside the waves, some of the layers dipping in the direction of the flood current and others with the ebb. The cross-bedding of marine sand waves is therefore more complicated in pattern than that found in river dunes.

Where the tidal currents are weakest a smooth sea bed appears underlain by patches or sheets of *mud and sand*. In places these deposits are 20 to 30m thick and fill up hollows, which had perhaps been scoured when sea-level stood lower.

Because of the greater strength of the currents, the sediments being deposited on the British shelf are appreciably coarser grained than those found in the Gulf of Mexico or off Nigeria.

Chapter 8

The Work of the Sea: the Ocean Deeps

Introduction
Our knowledge of the form, geology and processes of the deep sea is chiefly the result of work after the Second World War. This is not surprising, because the oceans, sometimes called 'inner space', are both vast and hard for men to explore. Their study became possible only because instruments were successfully made for probing an environment as hostile to man as outer space itself. Several of these—precision echo-sounders, sediment grabs, and coring tubes—were mentioned in connection with the continental shelves. Others helpful in studying the deep sea should now be described.

Learning about the deep sea
One of the most important of the newer instruments is called the *seismic profiler*, which works on the same principle as the echo-sounder. However, the profiler uses a different source of sound, usually either a high-voltage electric spark made in the sea or the explosion of a large bubble of combustible gas. As the result, powerful sound waves are produced which are able to penetrate through the bottom sediment and be reflected from hard or sandy layers tens or hundreds of metres below the sea bed. The reflected waves are received on board ship and their time of travel through the bottom sediment is changed by instruments into a vertical profile showing the sea bed and the structure of the layering in the underlying deposits.

Several instruments are used to find the temperature and salinity of ocean water. These include *thermometers* and *pressure gauges* to measure the change of temperature with depth, and *samplers*, varying in size from bottles to barrels, in which to collect sea water from different depths. Hence the distribution of the different layers of water can be found. Also useful is a *light-meter* (nephelometer) to measure the haziness of the water and therefore the amount of suspended matter. The slow movement of the different layers of ocean water can be followed by sinking into them a *neutrally buoyant float* fitted with a sound transmitter (pinger) so that its travel may be plotted. Surface currents can be tracked using *surface floats* or *drifting bottles*.

Underwater photography is now highly developed, and stereoscopic

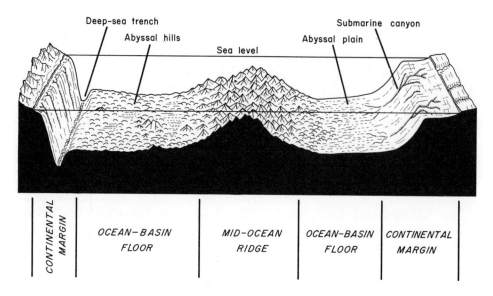

Figure 8.1 A model illustrating the chief morphological features of the ocean floor.

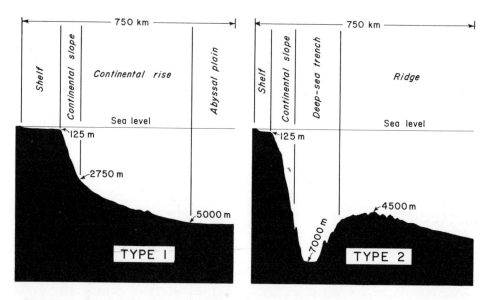

Figure 8.2 Vertically exaggerated profiles across two contrasted types of continental margin.

photographs as well as still or moving pictures can all be obtained. The camera lies in a strong water-proof housing mounted together with floodlights either on a sled dragged over the bed, or on a large tripod lowered from the surface. The photographs tell oceanographers a great deal about both the kind of sediment on the sea bed and the animals living there. Another important development of recent years is the *submersible*, a freely moving submarine allowing a crew of a few men to descend to great depths and closely examine the sea bed.

Continental margins
Earlier (Figure 1.3) the ocean floor was divided morphologically between: (a) continental margin, (b) ocean-basin floor, and (c) mid-ocean ridge. Figure 8.1 summarises in further detail what we know of this part of the face of the Earth.

The character of the outer part of the continental margin varies from place to place (Figure 8.2). The continental shelf, the shallowest part of this unit, was described in the preceding chapter. In the Atlantic Ocean, most of the Indian Ocean, and small parts of the Pacific Ocean, the continental margin consists of (a) *shelf*, (b) *continental slope*, and (c) *continental rise*. By contrast, around most of the Pacific Ocean and some parts of the Indian Ocean, the sequence is (a) *shelf* (usually very narrow), (b) *continental slope*, (c) *deep-sea trench*, and (d) broad *ridge*.

The *continental slopes* are steep compared with most parts of the sea bed, having gradients between 1:6 and 1:40. Almost everywhere they are crossed by valleys leading from the shelf down to the ocean floor. The smallest valleys measure no more than tens of metres wide and deep. The largest, called *submarine canyons*, are majestic and may be a hundred or more kilometres long. At a depth of 2000–3000m the sea bed becomes gentler in slope, and the continental slope gives way to the apron-like *continental rise*, with a gradient of about 1:300. The continental rises are much smoother than the

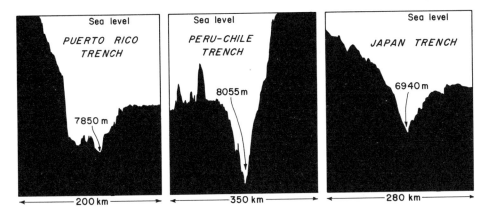

Figure 8.3 Some deep-sea trenches in profile. Vertical scales greatly exaggerated.

slopes, and their largest features are canyons crossing them every few tens of kilometres.

The *deep-sea trenches* are the deepest parts of the ocean, where the greatest water depth ranges between 6700 and 11 000m (Figure 8.3). They form a broken series of curved moats along the eastern, northern and western shores of the Pacific Ocean. A large trench lies in the Indian Ocean, south of Java, and a small one occurs north of the Caribbean island of Puerto Rico in the Atlantic Ocean. Trenches are steep-sided and seldom wider than 50km from rim to rim, though they may be many hundreds of kilometres long. Most are V-shaped in cross-section with a smooth flat floor, called a *trench abyssal plain*, a few kilometres across. Explorations made using the seismic profiler reveal thick sediments beneath the trench abyssal plains.

In mountainous parts of the land, one often finds that the rivers, flowing in steep canyons, pass down to gently sloping alluvial fans (for example, Figure 5.4). Surprisingly, a similar association of *canyons and fans* occurs in the oceans. The deep and majestic submarine canyons crossing the continental margins join the shallow shelves with the deep ocean-basin floor (Figure 8.4). The canyon heads, commonly with tree-like tributaries, generally

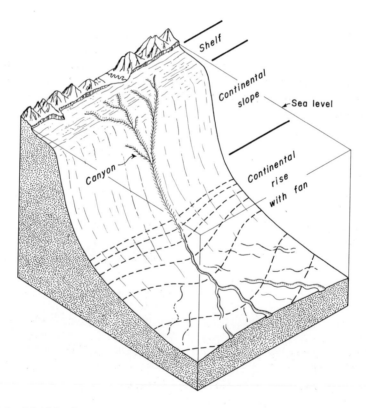

Figure 8.4 Model of a submarine canyon and associated deep-sea fan. Vertical scale greatly exaggerated.

lie opposite a large river or deep valley on the land. The canyons are tens or hundreds of metres deep where they cross the shelf and the continental slope, and commonly receive large tributary valleys. Their walls are steep and in places nearly vertical or overhanging; *rock outcrops* have been observed on these walls. At the bottom of the continental slope, where the sea bed grows less steep, the canyons empty on to large smooth spreads of sediment known as *deep-sea fans*. These correspond to the alluvial fans familiar from the land. Like alluvial fans, they slope gently away from the canyon mouths and show radial distributary channels. These channels are commonly bordered by sediment ridges corresponding to the levees of rivers.

Ocean-basin floor

The ocean-basin floor lies seaward of the continental margin and consists of two main parts, the *abyssal plains*, and the regions of *abyssal hills*.

The *abyssal plains* are flat smooth areas where the water depth is about 5000m and the gradient of the floor lies between 1:1000 and 1:10 000. They vary greatly in area, from the smallest of about one thousand square kilometres to the largest of approximately one million square kilometres. The monotonous flatness of abyssal plains is disturbed only by isolated hills and mountains and by occasional flat-bottomed deep-sea channels.

Every ocean has abyssal plains. Strings of them lie on either side of the Atlantic Ocean (Figure 8.5), and again along the western shores of the Indian Ocean. A very large abyssal plain occurs in the Bay of Bengal, south of the River Ganges. Some of the largest abyssal plains occur near the Aleutian Islands and Alaska. The archipelagoes of volcanic islands in the central and western Pacific Ocean overlook abyssal plains.

The *abyssal hills* generally lie further from the continents than the abyssal plains and often in slightly deeper water (Figure 8.1). In shape the hills are smooth and rounded (Figure 8.6). They vary in height from several metres to several hundred metres, and between 200 to 300m and 2 or 3km in width. The hills bear a thin cover of sediment, in contrast to the much thicker sediments underlying the nearby abyssal plains. Abyssal hills form large regions in all the oceans and they are more extensive than the abyssal plains. In the Pacific Ocean, for instance, abyssal hills cover especially large areas, particularly on the broad up-domed areas of the ocean bed known as *submarine rises*.

Mid-ocean ridge

The *mid-ocean ridge* (Figure 8.1) is an enormous world-encircling chain of submarine mountains formed of basaltic rocks. It has a total length of about 60 000km and is in many places about 1000km wide. Hence these underwater mountains rival in scale the great mountain chains of the land (Figure 8.7). Along the ridge crest their peaks stand 1000 to 4000m above the nearby abyssal plains and hills, the tallest reaching to the surface as islands. The topography of the ridge is commonly rugged, with deep valleys and many steep debris-strewn slopes and scarps.

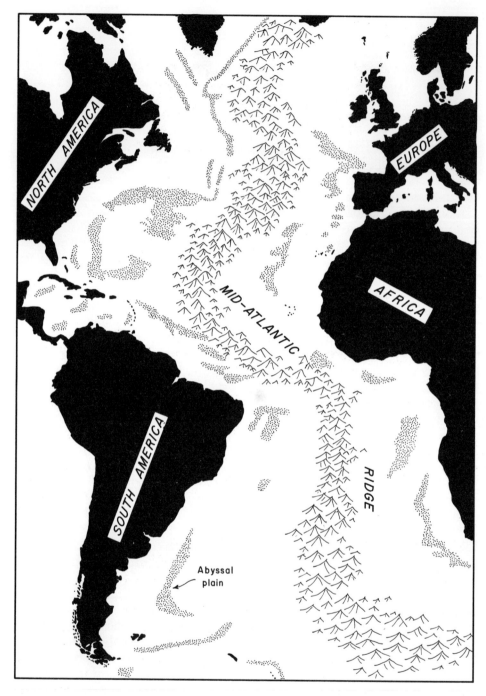

Figure 8.5 Abyssal plains flank the mid-ocean ridge in the Atlantic ocean.

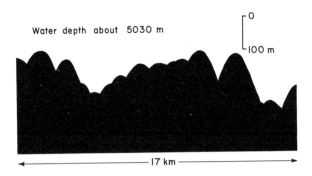

Figure 8.6 Abyssal hills on the floor of the east-central Atlantic Ocean. Vertical scale of the profile greatly exaggerated.

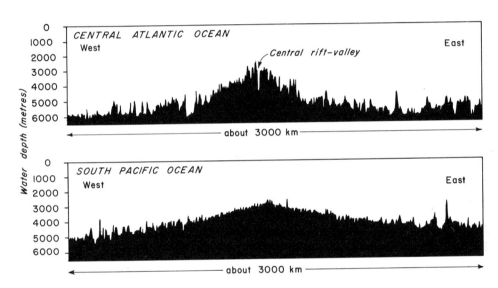

Figure 8.7 Two greatly exaggerated vertical profiles across the mid-ocean ridge.

Each ocean has a mid-ocean ridge lying along its centre. This is very clear in the Atlantic Ocean, where the ridge crest follows almost exactly the middle of that ocean and closely resembles in plan the coastlines of the Americas to the west and Europe–Africa to the east (Figure 1.3). The ridge in fact crosses the Arctic Ocean from the Siberian continental shelf before entering the Atlantic. In the Indian Ocean the ridge divides, one branch extending north to the Gulf of Aden, and another passing between Australia and Antarctica with an offshoot to New Zealand. The mid-ocean ridge circles the southern Pacific Ocean and ends up along the west coast of North America.

The mid-ocean ridges are in many places cut across by groups of *very large faults* or *fracture zones* which offset one section horizontally in relation to adjacent sections. These fractures are very numerous in the Atlantic and western Pacific Oceans, where they appear on the sea bed as great *submerged escarpments* or *ridges* hundreds of kilometres long. There is evidence, too, of fractures that lie *parallel* to the ridge crests. In the Atlantic Ocean (Figure 8.7) the crest contains a narrow but deep valley following the line of the ridge. It was formed when the rocks of the ridge were let down in a narrow block between two large parallel faults, the fault planes forming the valley walls. This central depression in the ridge crest is called the *central rift-valley*. Its floor is locally 1000 to 2000m below the peaks of the nearby submarine mountains and in places is formed of smooth sediment.

Sedimentary processes in the deep sea

Most sediment deposited in the oceans originates on the *continents*, from which it is carried by winds, rivers, and glaciers. Small amounts come from the weathering of rocks exposed on submarine hills and mountains, and a tiny quantity reaches the oceans from space in the form of meteoritic dust. This consists of small spherules of nickel-iron or silicate minerals. The organisms living in the oceans, notably the *plankton*, also contribute to the bottom sediments.

Winds blowing off the land carry large amounts of *mineral dust* to the ocean. This we know from both the composition of deep-sea sediments and from the *seasonal haziness* of the air in many parts of the world (Figure 8.8). The Trade Winds bear huge quantities of dust from the Sahara and spread it over the Atlantic Ocean, often as far west as Barbados. The Arabian Desert is also very productive, shedding much dust into the western Indian Ocean. Haze due to wind-blown particles also appears off the arid coast of Chile and Peru and off the coast of California. After settling from the air on to the ocean surface, the dust gradually sinks to the sea bed. In settling through the water it is further dispersed by the oceanic currents. From time to time *volcanic eruptions* introduce large amounts of dust into the atmosphere. The dust eventually settles on the ocean floor, where it commonly forms widespread layers of a distinctive composition.

Where glaciers end in the sea, as in Antarctica and Greenland and on the shores of the North Pacific, they break up to form ice-bergs. These drift away in the ocean currents and may travel for hundreds of kilometres before completely melting (Figure 8.8). Hence stones, sand and mud caught in the ice become widely dispersed over the ocean floor. They are said to be *ice-rafted*.

Rivers entering the sea deposit sand and stones close to shore. Mud is deposited over the outer continental shelves and some way beyond. Only the smallest particles remain suspended long enough to be carried directly by currents to the ocean deeps. There are, however, two ways in which sediments first laid down on the shelves can reach deep water, that is, undergo *resedimentation*.

The high winds that accompany storms commonly pile up water against

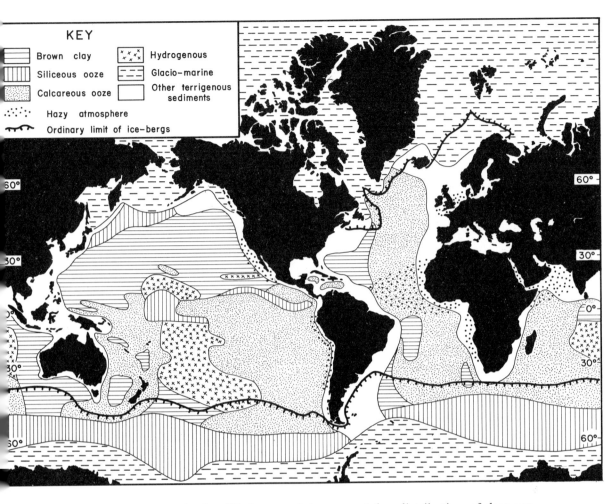

Figure 8.8 A greatly simplified map of the present-day distribution of deep-sea sediments.

the land and make waves strong enough to scour up mud and sand at depths of many tens of metres. The muddied water so formed on the shelves is more dense than the ocean water, and therefore spills down into the ocean basins. The outward movement is helped by the fact that the water heaped against the shore escapes by flowing out to sea as a bottom current (see Figure 7.3). Hence along most continental margins there is a zone of muddied water several hundred metres thick from which sediment (chiefly mud) stirred up on the shelves is redeposited. These zones are called *nepheloid layers*.

The second resedimentation process involves *turbidity currents* as transporting and depositing agents (Figure 8.9). The sediments on the steep slopes near canyon heads and along the shelf edges are usually unstable, because of their loose packing. Hence the shock of an earthquake, or a sudden violent storm,

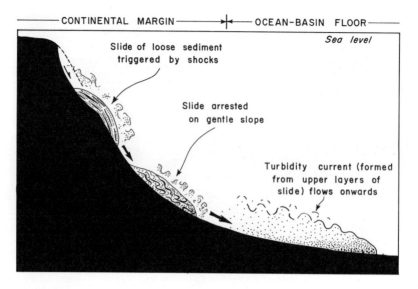

Figure 8.9 A model of the origin of turbidity currents.

is often sufficient to dislodge them. A huge mass of sediment will then move off downslope under the pull of gravity. The better consolidated lower parts of the mass are likely to remain coherent and so travel as an enormous *slide* of eventually contorted beds. The weak upper layers, however, will be diluted with sea water to form a slurry or turbidity current, similar to the one made in the bath (see Figure 3.4). The current will be *channelled* down any nearby submarine canyon and may travel for tens or hundreds of kilometres over the sea floor. In this way mud, sand and gravel first deposited in shallow water can be carried far from land into great depths.

Several recent turbidity currents have been studied. The best known followed the earthquake of 1929 on the Grand Banks of Newfoundland, where numerous telegraph cables cross the ocean bed (Figure 8.10). Several cables near the epicentre of the earthquake were severed by a gigantic slide which immediately followed the shocks. The turbidity current formed from the slide broke the other cables in outward succession, starting with the shallowest closest to the epicentre. The timing and spacing of these breaks showed that the speed of the current measured tens of metres per second at the start. Even after a distance of travel of 600km, it was flowing at a speed of several metres per second!

Turbidity currents on a similar scale to this are thought to occur frequently in many parts of the ocean. They appear to be responsible for the erosion of submarine canyons and for much of the deposition on deep-sea fans and abyssal plains.

Plankton consists of tiny animals and plants living in the surface waters of the oceans. These organisms make their hard parts by extracting from sea water the dissolved substances brought in by rivers. The most important of the

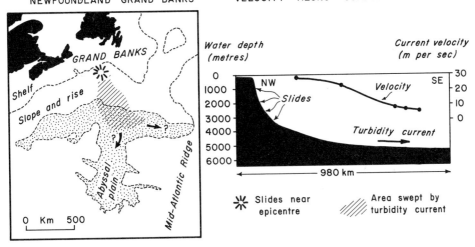

Figure 8.10 The scene of the Grand Banks earthquake and turbidity current of 1929, and some features of the current.

planktonic organisms are *foraminifera, pteropods, radiolaria, diatoms* and *coccoliths*. The foraminifera build chambered calcareous shells. A trumpet-shaped calcareous shell is built by the pteropods, cousins of the familiar snails. Radiolaria, diatoms and coccoliths are simple plants. Coccoliths are algae that cover themselves on the outside with small wheel-like calcareous plates. Diatoms and radiolaria, however, build beautifully ornamented shells of opaline silica.

When these animals die their hard parts sink gradually towards the ocean bed, entering regions of increasing water pressure and decreasing temperature. Calcareous shells become *more soluble* in sea water with increasing pressure, and so organic debris of this composition tends to dissolve as it sinks through the ocean. Because of this effect, deposits of calcareous shells do not form where the sea bed is deeper than about 5000m, even though the waters above are rich in plankton. The depth at which solution begins is, however, somewhat less in the cold polar waters than in warmer seas nearer the equator. In contrast, the siliceous shells of the diatoms and radiolaria are unaffected by increasing pressure, and they can accumulate whatever the depth of the sea floor.

The dissolved substances found in ocean water help to make deep-sea sediments in another way. Under certain conditions they can be chemically precipitated on the sea bed, or caused to react with minerals already deposited. The commonest of these chemically formed or *hydrogenous* minerals are rich in iron and manganese, though also containing noticeable amounts of other metals, for instance, copper and cobalt. They are deposited where oxidising conditions prevail at the sea bed, and form rounded *nodules*, *slabs* or *crusts*. The manganese nodules of the ocean bed are potentially an important source of raw material for man's use, and may one day be mined. Several other minerals are known to be precipitated on the ocean floor, including a number rich in phosphorous, another important raw material.

Distribution of deep-sea sediments
We now know enough about deep-sea sediments to show their distribution on a map (Figure 8.8). The main types are:

PELAGIC	TERRIGENOUS	HYDROGENEOUS
Brown clay	Mud	Manganese nodules,
Calcareous ooze	Turbidites	slabs and crusts
Siliceous ooze	Glacio-marine	

Figure 8.11 is a series of photographs illustrating the character of the sea bed where some of these kinds of sediment are to be found.

The *pelagic sediments* settled from suspension. The *brown clays*, widely formed in the Pacific Ocean, are exceptionally fine grained and consist of particles introduced by the wind and to a lesser extent first by rivers. They accumulate very slowly, at rates of about 1mm per century or less. As can be seen, the oozes are the most widely distributed of all deep-sea sediments. They are commonest on the abyssal hills and the mid-ocean ridge, but are in places unable to cover the underlying basalts completely. The *siliceous oozes* are commonest in high latitudes, whereas the *calcareous oozes* are best developed in temperate and equatorial regions.

The *terrigenous sediments* are coarser grained than the pelagic brown clays and contain noticeable amounts of silt. The *muds* are found close to the continents, mainly on the continental margin and the abyssal plains, where they may have been deposited from nepheloid layers. In the same regions *turbidites* are found. These are thin but horizontally extensive layers of sand deposited from turbidity currents. The typical turbidite bed is a few decimetres in thickness and rests on an irregular surface scoured by the turbidity current. The layer is graded from coarse and pebbly near the base to fine grained and muddy at the top. Turbidite beds are laid down very rapidly, at the rate of centimetres or decimetres per hour, in sharp contrast to the slowly accumulating brown clays, the oozes, and even the terrigenous muds. Studies with the seismic profiler have shown that the turbidites laid down on the continental rise and the abyssal plains are relatively thick and bury a topography similar to the abyssal hills.

The *glacio-marine sediments* occur mainly in a belt surrounding Antarctica. They are muds containing scattered to abundant sand and stones, and are more fully discussed in the next chapter.

Although *hydrogenous materials* contribute to many kinds of deep-sea

Figure 8.11A Pillows of congealed volcanic lava on the flanks of the mid-ocean ridge in the Indian Ocean. The pillows have a manganese encrustation and lie in 3870m of water. The bottom of the picture is 3m wide.

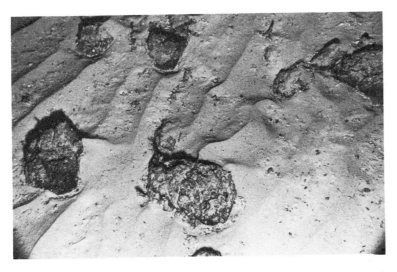

Figure 8.11B Scattered boulders on the sides of a seamount in the north-eastern Atlantic Ocean. The calcareous sand between the boulders has been shaped into ripples by the strong bottom currents, even though the water is 3127m deep. The bottom of the picture is 1.5m wide.

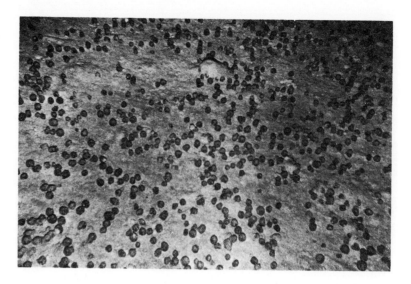

Figure 8.11C Uniform and rounded manganese nodules covering the sediment surface on the flanks of the mid-ocean ridge in the Indian Ocean. Water depth is 4230m. The bottom of the picture is 3m wide.

Figure 8.11D Sea bed of calcareous ooze in 3500m of water in the Indian Ocean. Many organisms have pitted or marked the bed while living there (see the spiral tracks). The bottom of the picture is 3m wide.

Figure 8.11E Calcareous ooze on the floor of an abyssal plain in the north-east Atlantic Ocean. Animals living in the bottom have left many pits and tracks. Water depth is 5340m. The bottom of the picture is 1.5m wide.

Figure 8.11F The flanks of a submarine ridge in the north-east Atlantic Ocean Water depth is 3391m. Steeply dipping sedimentary layers outcrop and the steep surface is partly covered with boulders and rubble. An incomplete mantle of calcareous ooze is present. The bottom of the picture is 3m wide.

sediment, they are most abundant in the central Pacific Ocean, where the extreme distance from the land results in an exceptionally low rate of sediment deposition. In this region the typical sediment consists of *manganese nodules* or *crusts* associated with brown clay.

Chapter 9
The Work of Ice

Introduction

We saw in Chapter 4 that a small but significant part of the Earth's moisture is at present stored in the form of *ice*, which occurs in cold lands as large, variously shaped masses known as *glaciers*. Some glaciers are like rivers, partly filling valleys, whereas others cover the land as huge sheets. Glaciers are in constant downhill motion under their own weight, as fresh snow is added to them on high ground and ice melts along their low-lying snouts. This flowing ice attacks the rock at the glacier bed, tearing and grinding it away. Thus the glacier carves the rock into distinctive shapes, which appear only when the ice finally melts. The debris the glacier so entrains is carried to far distant places and, when the ice wanes, is left as series of distinctive landforms and deposits. Some glaciers, however, empty into the sea. Drifting ice-bergs calved from them carry debris far and wide over the oceans.

Geologists know that at several times during the past few million years—the Quaternary era—the Earth's glaciers were much thicker and more extensive than today. The Quaternary era is in fact an *ice age*, a time when glaciers are unusually large and wax and wane abundantly. We sometimes call it *the* Ice Age, but this is perhaps unfair to the several earlier ice ages (e.g. the Permo-Carboniferous glaciation of the southern continents).

There are many *theories* about ice ages. Some scientists think they occur only when continents wander over the poles. Others suggest that polar regions may cool down drastically if ocean currents change their pattern, so that the heat supply from equatorial areas is reduced. Several scientists think that an ice age will occur if movements within the Earth cause the land to be uplifted too much. Some again believe that ice ages are produced when the heat energy received by the Earth from the Sun becomes diminished. A number think that changes in the Earth's orbit, or in the attitude of the axis of rotation, may cause the necessary cooling. Others suppose that an ice age would arise whenever the amount of carbon dioxide in the Earth's atmosphere was reduced. Although it is hard to choose between these ideas, there is no doubt that glaciation was frequent in the history of the Earth. Very probably the Earth has always had some glaciers.

Occurrence and shape of glaciers

Glaciers occur in *high latitudes* and at *high altitudes*, wherever it is cold enough for snow to lie on the ground throughout the year, and so in time change into ice. Near the poles, glaciers are found at sea-level, as in Antarctica, Greenland and Spitsbergen. In equatorial latitudes, however, glaciers (invariably small) can only be found 4000 to 6000m up the mountains.

The snow freshly deposited on a glacier is light and fluffy and consists of *loosely arranged crystals* associated with large amounts of entrapped air. Gradually, as more snow accumulates above to add its weight, the flakes recrystallise into larger, firmer and more closely packed grains. After this process of *recrystallisation and compaction* has gone on for many years, the buried snow is found to have become glacier ice, a *dense polycrystalline substance* containing only a few air bubbles. The ice nevertheless remains *stratified*, each layer, commonly marked by wind-blown debris, representing a winter's snowfall. It is difficult to believe that this hard substance, when part of a glacier, can flow plastically under the weight of the whole. This seems to be possible because different parts of each ice crystal can slip along internal planes of weakness.

Glaciers have many shapes and sizes (Figure 9.1). The largest are *ice-sheets*, of which two are known today. The Antarctic ice-sheet, with an area of $11.5 \times 10^6 km^2$ and an average thickness close to 2500m, has 85 per cent of the

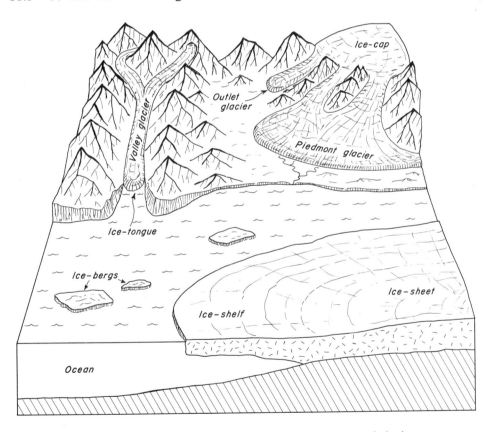

Figure 9.1   A model showing the occurrence of different types of glacier.

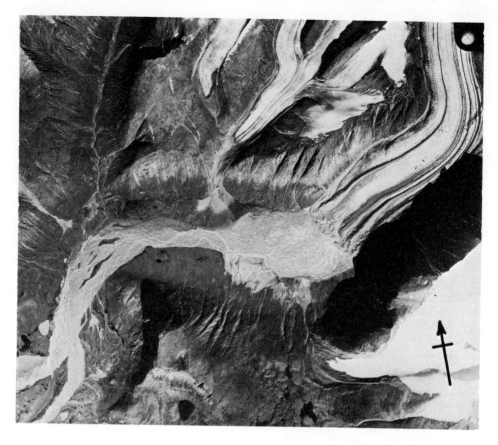

Figure 9.2A Vertical air photograph showing outlet (valley) glaciers of the Penny Icecap, Baffin Island, Canadian Arctic. The area measures approximately 13 ×11km. The glaciers are crevassed and show median and lateral moraines. Sediment carried from them by meltwater is forming a braided valley-train.

Earth's ice. The second is the Greenland ice-sheet, only one-eighth as large as the Antarctic ice-sheet. The next largest glaciers are *ice-caps*, examples being in Scandinavia, Iceland, the Canadian Arctic, and South America. In mountainous areas like the Alps one finds numerous *valley glaciers* (Figure 3.6). These are rivers of ice which, forming high up in the snow fields, use mountain valleys as channels to reach lower ground. Ice-sheets and ice-caps perched on high ground are commonly drained by a series of *marginal valley glaciers*, also called *outlet glaciers* (Figure 9.2A). Some valley glaciers flow to the sea, over which they spread as floating *ice-tongues*. A floating *ice-shelf* forms where several outlet glaciers join up on the surface of the ocean. There are several ice-shelves bordering Antarctica. Outlet glaciers that join together on lowlands below mountains form what is called a *piedmont glacier*.

Figure 9.2B Glacial striae on an outcrop of slate, Islay (Argyll), Scotland.

Figure 9.2C Glen Rosa, Isle of Arran, looking north towards Cir Mhor. Compare this U-shaped valley, now empty of ice, with Figure 3.6.

Figure 9.2D The glacial till shown here, exposed near Irvinstown, Fermanagh (Northern Ireland), is unbedded and very poorly sorted.

The coldness of glaciers

The way glaciers behave seems to depend very much on their *coldness*. Where the climate is especially severe, or where snow is added rapidly, the ice is very cold and the moving glacier remains permanently frozen to its rocky bed. The summer's meltwater is then confined to streams crossing the upper surface of the ice. Such glaciers are called *cold* or *polar*. By contrast, in areas of less severe climate, or where snowfalls are light, the ice commonly has a temperature very close to the melting point. Meltwater can occur throughout these glaciers, at the rocky bed, and in crevasses and tunnels, as well as on the upper surface. These glaciers are known as *warm* or *temperate*. They flow faster than polar glaciers, because the pervasive meltwater acts as a lubricant permitting basal slip, and they erode their beds more extensively. Figure 9.3 shows diagrammatically the pattern of flow in a valley glacier, an example of the temperate type. Glaciers nevertheless travel exceedingly slowly when compared with rivers or wind. The velocity of a glacier can generally be measured in terms of metres or tens of metres per year.

Glacier erosion

Quarrying is the process whereby a glacier removes large pieces and blocks of rock from its bed (Figure 9.4). It is possible because the bed-rock is ordinarily *jointed* or *fractured*, and because the melting point of ice is *lowered* by increase of pressure. Where a glacier flows over a rock hummock, the pressure at the base of the ice is increased on the upstream side, causing some

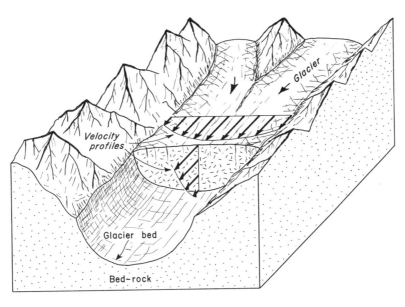

Figure 9.3 A model of flow in a valley glacier as shown by profiles of ice velocity. Part of the glacier has been cut away to show the cross-sectional shape and the glacier bed.

ice to melt. On the downstream side, the pressure is relatively low, and here meltwater may freeze. The result is that blocks of rock from the downstream side of the hummock become firmly lodged in the bottom of the glacier, which slowly carries them away. A characteristic shape is thus given downstream to the hummock, examples of which are often called *roches moutonées*, after a style of eighteenth-century wig.

Through quarrying a glacier becomes armed with debris that can be used for *abrasion* or *grinding*, as on the smoothly rounded upstream slopes of the *roche moutonée* in Figure 9.4. A glacier abrades its bed, rounding off irregularities, much as a carpenter smooths wood using sandpaper. Each piece of debris at the base of the moving ice chisels a *scratch* or *groove* into the rock, according to its size and hardness. These *striae* (Figure 9.2B) occur in large numbers on *glaciated rock surfaces* and lie parallel to the direction of ice flow. Their directions can be mapped to tell us the paths of long-vanished glaciers (Figure 9.5). The materials produced as the result of glacier abrasion, and also incorporated into the ice, are finely powdered but chemically unaltered minerals called *rock flour*.

The largest features eroded by glaciers are the troughs and valleys that many of them occupy. *Glaciated valleys* (Figure 9.2C) have a characteristic U-shaped cross-sectional profile. The walls are generally steep and locally almost vertical. Commonly the longitudinal profile shows a series of alternately steep and gentle slopes, sometimes with closed basins. The glacier flows relatively swiftly down the steep slopes, often forming an *ice-fall*, but much more slowly through the basins. An ice-fall therefore resembles a waterfall on

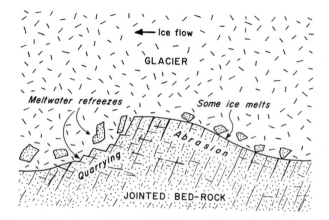

Figure 9.4 Model showing in longitudinal profile a glacier flowing over a rock hummock (*roche moutonnée*) on its bed. Abrasion is taking place on the upstream side of the hummock. Quarrying is occurring on the downstream side, where meltwater is refreezing in the joints and fractures of the rock.

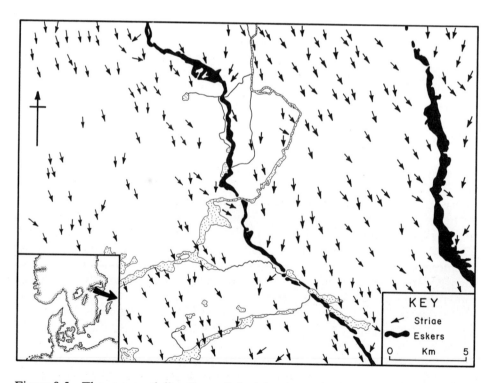

Figure 9.5 The measured directions of glacial striae and eskers in a small area 90km west of Stockholm, Sweden. Note the similar trends of the eskers and striae.

the course of a river. Many valley glaciers start high up in the mountains in what are called *cirques*, each of which in shape is a tilted basin, partly overlooked by steep eroding cliffs. The growth of cirques towards each other is in fact partly responsible for the sharp, jagged character of mountains in cold climates. Nevertheless the full shape of a cirque cannot be appreciated until after the ice has melted. Good examples of cirques can be found in mountain areas of Britain, such as Snowdonia, the Lake District, and the Grampians.

Glacial deposits formed on land

Some of the debris entrained by glaciers appears on the surface of the ice in the form of great dark-coloured trails known as *moraines*. In the case of fused valley glaciers, these moraines occur in *lateral* and *median* positions (Figure 9.2A). Much debris is also trapped within the ice and so concealed from view. It is left behind where the glacier melts after stagnating or reaching its terminus, to form deposits known as *glacial till*. Boulder clay is the older name for these deposits, but it is misleading, because boulders are not always present nor are the deposits invariably clays.

The typical till lacks internal bedding and is *very poorly sorted*, containing debris of a huge range of sizes, from tiny clay particles to great blocks of rock (Figure 9.2D). Two distinctive features may be shown by the larger pieces of debris. They may have one or more flat to slightly concave faces, or *facets*,

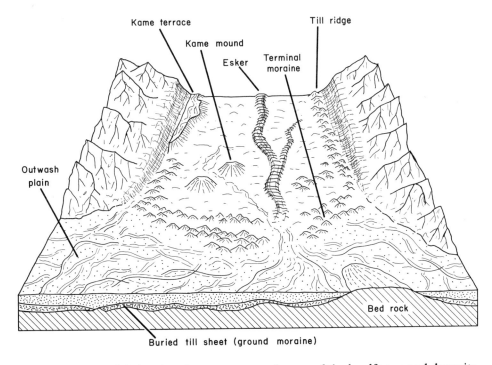

Figure 9.6 A model showing the occurrence of some of the landforms and deposits produced by, or in association with, glaciers.

which were formed as the stones rubbed against each other in the moving ice. Secondly, their surfaces may show delicate *striae* in different directions, another feature recording abrasion.

The median and lateral moraines of a valley glacier may give rise to long irregular *ridges of till* along the sides and in the centre of a valley (Figure 9.6). If the glacier remains stationary for a long time, the debris-charged ice constantly fed to the snout forms on melting a very irregular mound of till arranged across the valley. This is a *terminal moraine*, and several may be found in a single valley, showing that the snout paused occasionally.

Ground moraine or *lodgement till* is by far the most important kind of till, volumetrically and in area. It forms beneath slow-moving or stagnant ice which is melting at the base and usually occurs as a thin sheet-like layer incompletely covering a hummocky glaciated rock surface (Figure 9.6). Such a sheet may reach over tens of thousands of square kilometres, as in Scandinavia and the Canadian Lowlands. Some of these till layers have an irregular surface pock-marked by hollows that once contained huge abandoned blocks of ice. Other sheets of ground moraine have regular features on the surface. These may in places be long regularly spaced *ridges and hollows* either parallel or at right-angles to the direction of ice flow. Some sheets were moulded plastically by flowing ice into *drumlins* (Figure 9.7), which are rounded oval hills formed together in large numbers, giving a 'basket of eggs' landscape. Drumlin fields abound in the Scottish Lowlands and in Northern Ireland.

Figure 9.7 Drumlin landscape in County Down, Northern Ireland.

The stones in ground moraine commonly have their long axes arranged parallel to the direction of movement of the glacier that deposited the moraine. This *preferred orientation* of the stones, often called the *stone fabric*, is a feature inherited from the time when the debris was enclosed by ice and being incorporated into the till. If we measure in the field the orientation of a sufficient number of stones, by means of a compass and clinometer (dip-measurer), we can establish the stone fabric. This tells us, like glacial striae, the flow direction of the ancient glacier.

The moraines just described are deposited directly by the ice and accumulate in contact with the ice. Two other kinds of accumulation—*eskers* and *kames*—also form in contact with ice, but are due to meltwater and not glaciers directly (Figure 9.6). Such accumulations may be broadly described as *fluvioglacial*.

Eskers are long snake-like ridges often found with ground moraines, excellent examples occurring in Scandinavia (Figure 9.5). The largest measure hundreds of kilometres long, hundreds of metres wide, and tens of metres high. Usually eskers occur in groups having a similar trend, parallel to the direction of ice flow as shown by striae and stone fabrics. Commonly they show a tree-like branching, in the opposite direction to glacier flow. When eskers are quarried they are found to consist of well sorted and bedded gravels and sands showing many features, such as cross-bedding and delta structures, proving deposition from fast-flowing water. The cobbles and boulders forming the gravels are always well rounded, which would seldom be true if they occurred in a till.

Most eskers were formed inside stagnant ice by streams of meltwater flowing in *tunnels* within the glacier or upon its bed. Hence eskers can be seen only after the glaciers have melted, in the process of which some must have been let down onto the glaciated surface or ground moraine. It is not surprising that eskers in plan look like river systems and have similar sediments to fast rivers. Eskers are in fact the record of an englacial or subglacial system of drainage.

Kames also are formed by meltwater in contact with ice, some being steep-sided *mounds* and others short *winding ridges* (Figure 9.6). Like eskers, most kames consist of gravel and sand, though muds are sometimes also found. A stream of meltwater flowing across the surface of the glacier may deposit a kame. Other kames originate through the infilling of crevasses or vertical shafts in the ice. Some may represent deltas built into lakes on top of the glacier.

Proglacial deposits

Beyond the glacier snout lies a region affected by *meltwater streams* and by the *wind*. This is the *proglacial region*, where *proglacial deposits* form.

The meltwater streams are heavily laden and occupy constantly shifting braided channels. If the glacier snout lies well up a valley, the streams lay down in the valley an outwash deposit known as a *valley-train* (Figure 9.2A). By contrast, an *outwash plain* underlain by outwash deposits is formed where a glacier empties on to extensive lowlands (Figure 9.8). Generally, outwash deposits are well sorted, rounded and bedded gravels and sands which fill up channels and show structures such as cross-bedding.

The braided streams shift so rapidly that plants can secure little hold in the

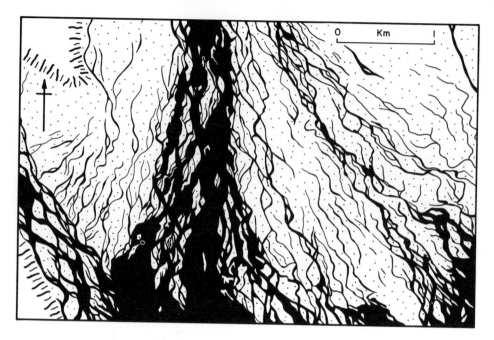

Figure 9.8 A part of the braided outwash plain south of the ice-cap Vatnajokull in Iceland.

outwash. Consequently the wind is free to scour the surface and to blow up sand and dust, just as in hot deserts. *Ventifacts* and *wind-blasted rocks* are quite common in glacier outwash plains, and wind-blown sand may accumulate locally as large well-shaped *dunes*. The dust is readily suspended and borne far away in great clouds to be deposited as a blanket of *loess*. This is a distinctive material, consisting of angular silt-size particles. Insects and land animals are commonly preserved in loess.

Glacial deposits at sea
Glaciers that reach the sea lose mass in two ways. They may *melt underneath* in contact with the warmer sea water, and they may break up into ice-bergs, a process called *calving*. Ice-bergs drift with winds and currents for hundreds of kilometres over the ocean before completely melting away. They commonly *run aground* in shallow water, and so may *groove* the sea bed on the continental shelves and upper slopes. Many of these grooves have now been detected using side-scan sonar (Figure 9.9). They are metres in depth, tens or even hundreds of metres wide, and many kilometres in length. If fossilised they would form important evidence for the action of ice. Some of Pleistocene age were recently found off the west coast of Britain.

The debris trapped in the ice is gradually released as the floating glaciers and ice-bergs melt, and is said to be *ice-rafted*. The loosened particles sink through the water to the sea bed, where they add to the bottom sediments. On

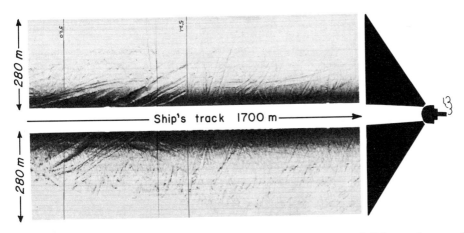

Figure 9.9 Double side-scan sonar record of grooves made by drifting and grounding modern ice-bergs on the shallow bed of the Beaufort Sea, Arctic Ocean. The grooves are hundreds of metres long, tens of metres wide, and metres deep.

Figure 9.10 In the Horlick Mountains, Antarctica, there are Permo-Carboniferous interbedded mudstones and sandstones containing stones (shown by arrows) rafted in by ice-bergs.

the sea bed beneath a floating ice-tongue or ice-shelf one finds a jumble of stones, sand and mud with the preserved remains of marine organisms. Further out to sea, in the zone of drifting ice-bergs, the sea floor consists of muds with scattered sand grains and large stones (Figure 9.10). The mud is rich in the remains of deep water marine animals yet contains striated stones dropped from the ice-bergs. The stones as they fell from above often disrupted the bedding in the mud, just as a brick thrown into soft mud splashes up and distorts the deposit. Sediments of these kinds cover tens of thousands of square kilometres of the sea bed in polar waters. They are therefore as important geologically as the lodgement tills found on the land.

Chapter 10

The Restless Lithosphere: Evidence from the Past

Introduction

The Earth's crust—the uppermost layer of the lithosphere—is easily studied by looking at the strata exposed in quarries, in stream beds, or on coastal cliffs. One of the impressive things about these rocks is their apparent rigidity and great strength. Yet in the same places we can often find evidence showing that the strata have been *deformed*, that the shape and attitude and relationship of the beds became changed in some way after the rocks were produced.

The *deformations* are of several kinds. In some cases the rocks were *broken across* or *faulted* along a surface of fracture, the broken ends of the beds being moved in relation to each other. In other instances the beds were *bent* or *folded* into wave-like or more complicated shapes. Often strata are faulted as well as folded. Some exposures show us examples of rocks which have been *stretched* or extended horizontally. Rocks which have been *compressed* horizontally can also be found.

These changes in the strata are all due to the action of various *forces* within the lithosphere during the course of geological time. *Gravity* acting on the mass of the rocks sets up forces that are often large enough to cause folding and stretching. It is important to remember that these *gravitational forces* are *vertical* to the surface of the Earth. Another set of forces acts *horizontally*, parallel to the surface of the Earth, and may be due to the action of huge but very slow *convection currents* within the mantle. As we shall see in more detail in Chapter 11, these currents appear to drag portions of the lithosphere sideways over the surface of the Earth. Hence they could *stretch* or *compress* rocks horizontally.

The various structures formed in rocks through the action of these forces are called *tectonic structures*. They include *folds, faults, joints, cleavage*, and many others, and are developed on various scales. The smallest folds and faults can be seen in a single hand-specimen. In order to detect the larger structures it may be necessary to map considerable areas geologically. The *orogenic fold-belts* are perhaps the largest tectonic structures of all. They are complicated assemblages of folded, faulted and often metamorphosed rocks which stretch in narrow belts for hundreds or even thousands of kilometres across several countries or even a whole continent, for instance, the Alpine fold-belt (Tertiary) shown in Figure 1.4.

The way rocks deform
When rocks in the lithosphere are deformed they obey the same physical laws that govern the deformation of other solid substances such as metals and plastics. Some of these laws can be illustrated by simple experiments.

One experiment is designed to show how a long vertically hanging wire behaves when loaded with weights. One finds that the wire grows longer when weights are hung from it. Each weight represents a *force* available to deform the wire, whereas the amount by which the wire stretches records the *response* to the force. We can describe this response by calculating the quantity

$$\frac{\text{elongation of wire (amount of stretching)}}{\text{unstretched length of wire}}$$

which is called the *strain*. It bears a relation to the deforming force that is unique to the substance being studied, and is important in the study of tectonic structures.

Suppose the experiment was done using a soft copper wire, and we plotted a graph of weight hanging from the wire against strain (Figure 10.1A). Over the rising part of the curve the strain increases each time the weight is increased. In this region of weights the copper wire behaves *elastically*, that is, it returns to its unstretched length if we remove the weights. The upper limit of this range of behaviour is marked by a critical weight known as the *yield point*. Beyond this point over the horizontal portion of the curve, however, the strain increases without any increase of weight, until eventually the wire breaks. The wire now is said to behave *plastically*, and the stretching is *permanent* Many rocks, especially the weaker ones like shale and rock salt, behave plastically and so can be permanently stretched and bent by considerable amounts before breaking under the action of the forces in the Earth. Even granite and well-cemented sandstone can behave plastically if they are

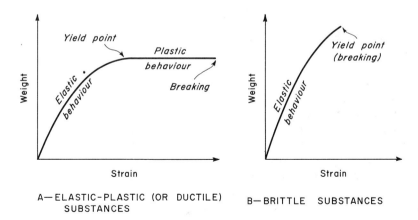

Figure 10.1 The behaviour of different kinds of substance when wires made of them are loaded with weights.

113

sufficiently softened by heat and pressure.

Some substances would not have behaved plastically in the experiment. If we had used a long glass fibre, our graph would resemble that in Figure 10.1B. The fibre *fractures* when a critical weight, defining its yield point, is reached. Glass and similar substances are called *brittle*, because there is no region of plastic behaviour following the region of elastic behaviour. Several kinds of rock behave in a brittle way, for instance, well-cemented sandstone, limestone, and granite, unless softened by heating.

The force in these experiments was a *pull* or *tension*. The deformation it caused can be represented by the *change of a square element* of the substance into a rectangle of the same area (Figure 10.2A). If we reverse the force in direction, making it a *push* or *compression*, we can again represent the deformation by the change of a small square into a rectangle (Figure 10.2B).

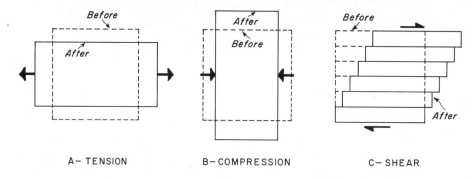

Figure 10.2 Different kinds of deformation, as illustrated by the change in shape of a square element.

Notice that with a compression the rectangle is longest at right-angles to the direction of the force. There is one more kind of rock-deforming force, a *shear stress* acting *parallel* to the surface on which it operates. This kind too can be represented by the change in shape of a small square of the substance. Imagine that the square is divided into a large number of slices parallel with one edge, like a pack of playing cards. When the topmost slice is disturbed, every slice below moves a little way, and the square is changed into a diamond shape (Figure 10.2C). The shearing therefore causes a *rotation* of two opposite sides of the square.

These three kinds of force—pulling, pushing, and shearing—generally act simultaneously in the Earth's crust. The deforming forces are therefore in reality complex.

Joints

Joints are extensive *cracks* to be found cutting most bodies of rock. They are of different kinds, but their presence invariably shows that the rock has behaved in a brittle way. Joints are in fact *brittle fractures*.

Figure 10.3 Well-jointed flat-lying Jurassic sedimentary rocks, near Leckhampton, Gloucestershire.

Figure 10.4 Polygonally jointed lava flow, Clam Shell Cave, Staffa (Argyll), Scotland.

In sedimentary rocks joints are regularly arranged planar cracks lying nearly perpendicular to the bedding (Figure 10.3). Usually they are aligned along two or three different compass directions. All of the joints aligned along each particular direction form a *set* of joints, whereas the joints of the several sets combine to form a *joint system*. The spacing of the joints in each set varies with the kind of rock and the thickness of the beds. Joints are also plentiful in igneous rocks and may lie far apart, though they are seldom regular. The joints just mentioned may have been caused either by a contraction of the rock or by forces that were not large enough to cause faulting.

In certain cases joints are definitely due to tension caused by the shrinkage of the rocks. Examples are the marvellously regular *polygonal joints* formed in igneous bodies, particularly dykes, sills and lava flows, as they contract on cooling. Splendid examples can be seen at the Giant's Causeway in Northern Ireland and at places in the Inner Hebrides (Figure 10.4).

Joints have much practical importance. They ease the work of quarrying but at the same time make mining and tunnelling dangerous jobs. The presence of joints commonly makes it easy for fluids such as natural gas and water to pass through rocks.

Faults

Faults are also cracks in the rocks due to brittle fracture. However, they differ from joints in two main ways. Faults are fractures ordinarily *much larger* and spaced *further apart* than joints, although the two kinds of structure often trend in the same direction. Secondly, the beds on either side of a fault are always moved in relation to each other, or *displaced*, by a distance varying between a few centimetres and many kilometres, according to circumstances. There is practically never an obvious displacement associated with joints.

Faults (Figure 10.5) are of three main kinds (a) *normal faults*, (b) *reverse*

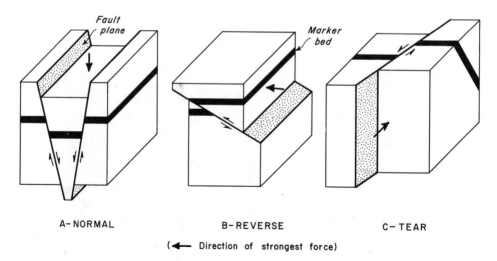

Figure 10.5 Different kinds of fault and their effects on marker beds.

faults, and (c) *tear* or *wrench* faults. Each kind is related to a *definite pattern of forces* acting in three-dimensions at right-angles in the Earth. A study of faults can help us to say whether the crust in a particular region has been *extended* (subjected to tension) or *shortened* (subjected to compression) because of these forces.

When *normal faults* (Figure 10.5A) are made the strongest force is a *vertical compression* arising from the weight of the rocks in the field of gravity. The other two forces, acting horizontally at right-angles, are weaker and one of them represents tension. The fault planes dip steeply and strike at right-angles to the direction of the tensional force. Faults of this kind lead to a *permanent extension* of the crust in the direction of the tension. They may also cause blocks of rock to drop downwards relative to adjacent blocks, forming large fault troughs on the Earth's surface. Each foundered block is called a *graben* and each upstanding one a *horst*.

Reverse faults (Figure 10.5B) represent the opposite arrangement of the forces. The strongest force is a *horizontal compression* while the least is the vertical compression arising from the weight of the beds. The fault planes strike at right-angles to the main force but may dip at various angles. Often a fault plane inclines gently to the horizontal, and sometimes actually lies parallel to the bedding. Such faults are called *thrusts* and they commonly represent a very large displacement. Reverse faulting leads to a *permanent shortening* of the crust in the direction of the compression. If the faults are of the thrust type, a great deal of shortening may have taken place.

The third kind of fault is the *tear* or *wrench* fault (Figure 10.5C). This kind forms when the strongest force is a compression in a horizontal direction and the weakest lies in the horizontal direction at right-angles. The downward-acting weight of the rocks is now the force of intermediate strength. The fault planes are vertical, with the movement on them horizontal. When tear-faulting occurs the crust is shortened parallel to the compression but extended horizontally parallel to the weakest force.

Normal faults are common in Britain, for instance, in the Carboniferous rocks of the Midland Valley of Scotland, where a system of faults composed of two sets is found (Figure 10.6). Very large normal faults with a displacement measured in kilometres are associated with the great *rift valleys* of the world (horst and graben structures), for example, the East African Rift.

The Carboniferous rocks of the Mendip Hills and the Avon-Somerset Coalfield show many reverse faults and thrusts (Figure 10.7). The Moine Thrust cutting the Precambrian and Cambrian rocks in north-west Scotland is another good example. Huge thrusts with displacements measured in tens of kilometres occur in the European Alps and the Appalachian and Rocky Mountains of North America.

Tear faults are very impressive. In Britain we have a good example in the Great Glen Fault in Scotland. By matching corresponding geological features on either side of this fault, it can be shown that the total horizontal displacement amounts to about 100km. The San Andreas Fault, stretching for about 1000km along the coast of California, is even more spectacular. The

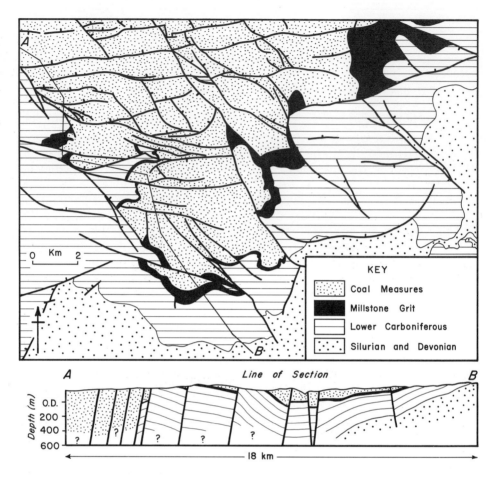

Figure 10.6 Pattern of normal faults in Carboniferous beds south-east of Hamilton, Lanarkshire, Scotland.

horizontal movement on this fault is at least 580km, and the fault is still active, as was shown during recent earthquakes. The rate of movement is on average about 5mm per year.

Deformed fossils and pebbles
When beds become jointed or faulted, the rock itself, as opposed to the strata, seldom experiences any permanent squeezing or stretching. Strata that are folded, however, often change greatly in thickness along their length because of the marked stretching and squeezing of the rock that goes with folding. The changes are greatest when the folding is strong and the rocks behave plastically. We can get a good idea of what happened by studying *changes in the shapes* of objects that formed a part of the rock, such as fossils, or pebbles, and originally spherical particles like ooliths. These objects serve, in fact, as *strain markers*.

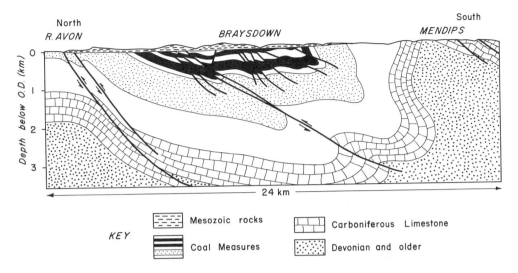

Figure 10.7 Reverse faults, thrusts, and overturned folds. A profile showing the geological structure between the Mendips and the River Avon east of Bristol.

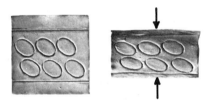

Figure 10.8 On the left is a Plasticine model of a pebble bed. The picture on the right shows the effect of compressing the bed at right angles to the bedding.

Pebbles are laid down with their large faces nearly parallel to the bedding, so that in a vertical section we would see them as ellipses. The Plasticine block in Figure 10.8 represents a gravel bed and the ellipses marked on it the pebbles. If the block is squeezed in a vice at right-angles to the bedding, we stretch the bed and make the ellipses (pebbles) even longer and thinner. *Stretched pebbles* are common in metamorphic rocks which, lying deep in the Earth, deformed plastically while soft at a high temperature and pressure. They show that the rock was *strained*. If we repeat the experiment with circles marked on the block, to represent spherical ooliths, we find that the circles become changed into ellipses, which grow longer and thinner as we increase the strain (Figure 10.9). Limestones with stretched ooliths can be found in the Alps.

Often in folded rocks are *fossils* deformed along with the enclosing matrix. A famous example is the trilobite *Angelina*, found in the Cambrian rocks of

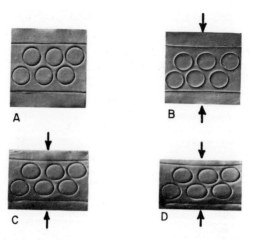

Figure 10.9 A Plasticine model of spherical ooliths is shown in A. Pictures B to D show how the ooliths became changed in shape as they were compressed more. Compare with Figure 10.8.

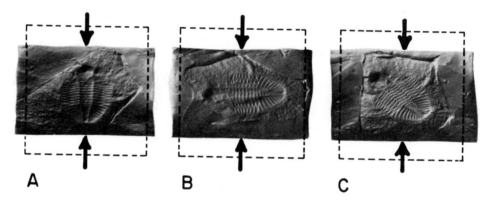

Figure 10.10 An impression of a trilobite has been made in three square blocks of Plasticine (dashed lines). The trilobite in A was shortened by compression, whereas the trilobite in B became lengthened. The trilobite in C was changed by compression from bilaterally symmetrical to unsymmetrical.

North Wales. What happened to this fossil can be shown by squeezing in a vice a block of Plasticine marked with the impression of a trilobite (Figure 10.10). The impression changes shape in different ways according to the chosen direction of squeezing. It may be either *shortened* or *lengthened*, or made to *lose* its bilateral *symmetry*.

Folds

Rocks bend and flow plastically when they become folded. Hard brittle rocks like sandstone and limestone yield to the forces causing folding partly by bending. They are called *competent*. The *incompetent* rocks—the weaker ones such as shale—respond mainly by flowing as if they were very viscous liquids.

We can easily find out what happens when a bed is folded. Make a long rectangular bar out of Plasticine and along one of the long faces mark two rows of circles. Now carefully bend the bar over the curved surface of a large bottle, keeping the marked face perpendicular to the glass. Many of the circles become changed into ellipses (Figure 10.11A). The ellipses in the *outer* row show by their direction of elongation that here the Plasticine was under tension. The ellipses of the *inner* row show that the Plasticine of the inner part of the bar was *compressed*. Another pattern of strain (Figure 10.11B), encountered in some folds, will be formed if you fold (buckle) the bar by pushing the ends closer.

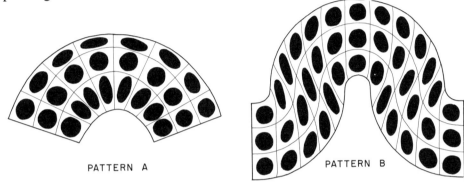

Figure 10.11   Contrasted patterns of strain in a folded layer.

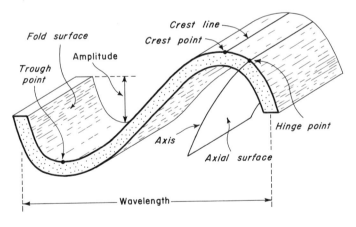

Figure 10.12   The chief features of a fold.

Folds are easily described (Figure 10.12). The surface separating any two adjacent layers of rock in a fold is called the *fold surface*. The highest point on the surface is called the *crest point* and the lowest the *trough point*. Between the trough and the crest is the *fold limb*, over which the fold surface changes

121

in direction of curvature. The fold *crest line* is formed by joining up the crest points in adjacent cross-sections. There are usually points on the fold surface where the radius of curvature of the surface is a minimum. Each of these points is called a *hinge point*. By joining hinge points on successive fold surfaces we can draw the fold *axis* and *axial surface*. Where there are several folds together, the distance separating two adjacent crests or troughs is the fold *wavelength*. The fold *amplitude* is half the perpendicular distance between a crest and a trough.

The *orientation* of folds is also important (Figure 10.13). Where the fold surfaces are *convex-upwards*, the fold is called an *anticline* (Figure 10.14). The commonest anticlines have the *oldest rocks* in the *core* of the fold, but included in the complex folds we shall call nappes (recumbent folds), there are anticlines with the oldest rocks towards the outside. The name *syncline* is applied to fold structures in which the fold surfaces are *concave-upwards*

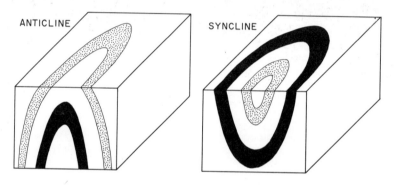

Figure 10.13 Anticline and syncline compared.

(Figure 10.15). The commonest synclines have the *youngest rocks* in the *core* of the fold. In nappes, however, you will commonly find synclines with the oldest rocks in this position.

There are certain varieties of fold which it is useful to distinguish (Figure 10.16). A *monoclinal fold* has a single limb joining flat-lying beds on either side. In *overturned folds* both limbs dip in the same direction but at different angles. A *dome* is an anticline with a roughly circular to oval outcrop in which the beds dip relatively evenly away from the core. The synclinal version of this structure is the *basin*, met in Chapter 4. A *nappe*, or *recumbent fold*, is a large complex fold with a flat-lying axial surface. Nappes commonly rest on huge thrusts and are typical of fold-belts.

Cleavage

Cleavage is a small scale tectonic structure often developed during strong folding (Figure 10.17). Cleaved rocks can be split into *thin plates* along parallel surfaces called *cleavage planes*, and they have been strongly com-

Figure 10.14 An anticline and a syncline in the Old Red Sandstone cliffs (40m high) near Dale, Dyfed, Wales.

Figure 10.15 The broad foreshore near Berwick-on-Tweed, Northumberland, shows a horizontal section across a syncline.

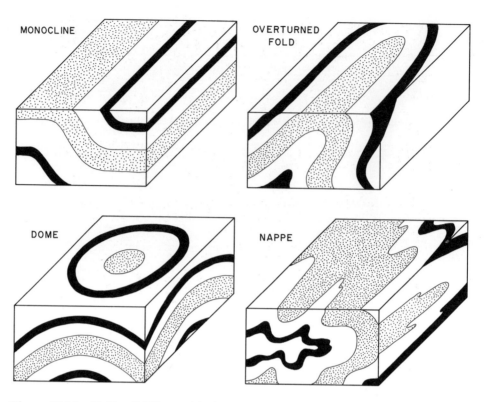

Figure 10.16 Folds of different kinds.

pressed along a roughly horizontal direction while deeply buried and subjected to warming beneath other rocks. The cleavage develops because, under these conditions, the mineral grains in the rock due to various causes become aligned at *right-angles to the compression*. Some grains are *mechanically rotated* into a sub-parallel relationship during the squeezing, whereas others are flattened. The micaceous minerals *recrystallise* to form new flaky crystals which lie perpendicular to the compression. It is the parallelism of the constituent mineral grains which allows cleaved rocks to be split so readily.

Cleavage is best developed in *slates*, which are deposited as muds rich in flaky clay minerals. The slates of North Wales are renowned as a roofing material, and good slates also occur in the Lake District. Sandstones, limestones and igneous rocks which contain few or no micaceous minerals never show a well developed cleavage.

Cleavage and folds have an interesting relationship, as you may see in Figure 10.14. Because the two develop roughly simultaneously, the cleavage surfaces in each fold have a definite attitude relative to the fold, forming a pattern known as a *cleavage fan*. Generally the cleavage in each fold is steeply inclined to the bedding and roughly parallel to the axial surface.

Figure 10.17 This sequence of alternating sandy and muddy rocks has become cleaved, the strongest cleavage occurring in the muddy bands. Note in Figure 10.14 how the cleavage is related to the folds.

Fold-belts

The largest of all tectonic structures are the *major fold-belts* or *orogens*. These are complexes of folded and faulted rocks, in parts highly metamorphosed, which extend in long narrow belts for great distances. In Europe there are three main fold-belts younger than Precambrian. The *Caledonian fold-belt* of Scandinavia and northern and western Britain is made of Lower Palaeozoic rocks. The *Hercynian orogen* was formed around the end of Carboniferous times and stretches east–west from central Europe into southern Britain. The youngest fold-belt is the *Alpine* which dates from the Tertiary. It is associated with the lands bordering the Mediterranean Sea. Each of these three fold-belts is represented in other parts of the world.

When fold-belts were formed the Earth's crust became very much shortened. In most fold-belts one can find huge thrusts and nappes besides a host of smaller folds and faults. Clearly fold-belts can only have been formed by the action inside the Earth of very powerful forces that affected regions on the scale of the present-day continents.

Chapter 11

The Restless Lithosphere: Evidence of Movement at the Present Day and in the Geologically Recent Past

Introduction

The last chapter described the variety of tectonic structures occurring in rocks which are amongst the oldest known from the Earth. These structures show that different parts of the Earth's crust (and, to complete the lithosphere, presumably also the upper part of the mantle) have moved relatively at various times in the past. In certain cases, the movement was essentially *vertical*, as when a rift valley appeared through normal faulting or a dome-shaped fold was formed. The movement in other instances was primarily *horizontal*, and often to the extent of tens or hundreds of kilometres, as witnessed by the great thrust and tear faults and by the nappes.

These structures are the fossilised proof that movement once upon a time occurred. We should ask ourselves whether there is any evidence that the lithosphere is moving today? It would be quite remarkable if the presence of Man upon the Earth corresponded to the time when the Earth ceased to alter, on account of a stoppage of its interior processes! It will be seen that the lithosphere has been moving in the geologically recent past, and continues to move at the present, commonly at a remarkably high speed.

Isostasy and isostatic equilibrium

Surveyors mapping the plains of India near the Himalayas found that when they had calculated the amount of rock present in these mountains, this attracted their plumb-bobs by smaller amounts than they had expected on the basis of the law of gravitational attraction between bodies. In order to explain this it was suggested that beneath the mountains lay deep 'roots' formed of low-density rock, and that the mountains and their roots 'floated' in a state of *hydrostatic equilibrium* on the more dense material around them. This state of equilibrium or balance is given the name *isostasy*. Hence any part of the Earth's crust which is in this state of balance is said to have *isostatic equilibrium*. The downward-acting weight of the part is then exactly equal to the upward-acting buoyancy force, by Archimedes' principle.

The idea of isostatic equilibrium is easily illustrated (Figure 11.1). Imagine that a continent with plains and mountain ranges is a series of adjacent blocks of wood of different thicknesses floating on water. In order that the blocks may be in isostatic equilibrium, there must be a greater thickness of wood under water in those representing mountains than in blocks acting as low-lying plains. We know that isostatic equilibrium occurs for very large portions of the crust of the Earth; it is only necessary to compare the relative position of the thick continental crust with much thinner crust under the oceans (Figure 1.3). It is also known that a close approach to isostatic equilibrium

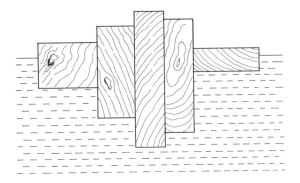

Figure 11.1 The idea of isostatic equilibrium. To be in equilibrium the blocks sink to depths which are a constant proportion of their heights.

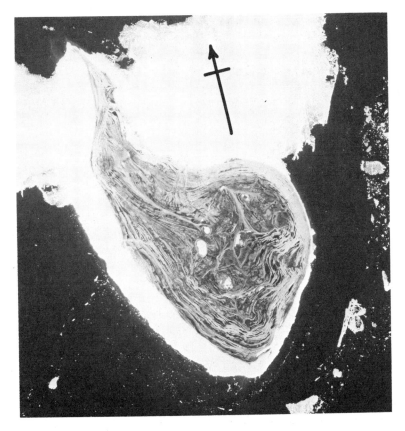

Figure 11.2 Air photograph of Tangle Island (8km long), Foxe Basin, Canadian Arctic. Encircling raised beaches can clearly be seen on the island, which is surrounded by ice and ice-bergs drifting in the dark sea.

occurs for the Earth's terrestrial and submarine mountain chains.

A ship rapidly sinks deeper into the water when cargo is loaded into it, as you can tell from the Plimsoll line painted on the side. When unloaded, however, the ship quickly rises some way out of the water. Similarly, a part of the Earth's crust should move up and down to maintain isostatic equilibrium if the weight of that part for some reason changes. Such *vertical movements* do occur, but not immediately the weight is altered. This is because the materials of the lithosphere are strong enough to resist for a while the movement required by the change of gravitational force.

Vertical movements representing *isostatic adjustments* are well known. During the Pleistocene vast regions in North America, Scandinavia, and northern Russia carried ice-sheets hundreds or thousands of metres thick. The ice formed a large additional weight on the crust, which was consequently pressed downwards, like the loaded ship. This weight was reduced as the ice melted, and the crust began slowly and jerkily to move upwards. These movements are still taking place. For instance, in Scandinavia and Canada there are *raised* marine *beach deposits* only 10 000 or so years old which now lie more than 100m above sea-level (Figure 11.2). They represent a maximum rate of upward vertical movement of the land comparable with one metre per century. Vertical movements representing isostatic adjustments should also occur as the continents lose mass through being weathered down. However, these movements are likely to be very gradual, because of the slowness of weathering. Similarly, sediment deposition should also cause isostatic adjustments, though these again are likely to be slow.

Continental drift

Raised beaches and the like show that the continents move vertically. Is there any evidence to show that they can drift, or *move horizontally* relative to each other, over the surface of the Earth? In fact, several lines of evidence strongly support the idea of *continental drift*.

A glance at an atlas shows that the east coast of South America matches in shape the west coast of Africa, suggesting that Africa and South America were at one time joined together but later drifted apart. This is an example of the evidence of the *topographic fit* of continents. The best evidence of this kind comes when the shapes of the continents are defined geophysically rather than in terms of modern coastline, and when the two shapes are matched by an electronic computer which can measure 'goodness of fit'. It will be remembered from Figure 1.3 that the true or geophysical margin of the continents lies at present in deep water, somewhere on the continental slope. As you can see (Figure 11.3), the fit of Africa and South America on this basis is very good, with no large gaps or overlaps. Comparably good fits have been found for North America with Greenland and Europe (Figure 11.4), and for Antarctica with Australia and Africa.

If continents have drifted apart, one should be able to *match the rocks* on either side of the presumed join as regards their types, fossils, ages and tectonic structures. In many cases a match of these kinds is possible. For instance,

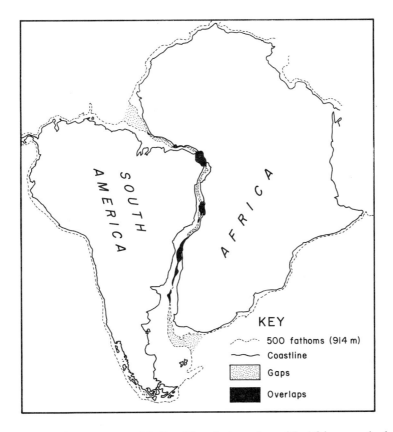

Figure 11.3 The best topographic fit of South America with Africa as calculated by a computer.

when North America, Greenland and Europe are fitted topographically, the great orogenic belts are found to line up (Figure 11.4). The Caledonian belt crossing Scandinavia, Greenland and Britain, continues into Newfoundland and the northern Appalachian Mountains. Similarly, the Hercynian belt curves smoothly from central Europe, through Britain, into North America. These similarities of regional geology are important evidence favouring continental drift.

Another way of testing the idea of drift is to map the *distribution of ancient climatic belts* using the character of the rocks of each age. You will be aware from earlier chapters that some kinds of sediment are restricted to certain major climatic zones. Glaciers, for instance, are active chiefly in polar regions. Therefore if evidence of glacial conditions occurs widely in the rocks at some place, we can reasonably suggest that those rocks were deposited near one of the Earth's geographic poles. Again, in the hot deserts which lie towards the equator, there are large areas deeply covered with sand blown into high dunes by the wind. Where similar but older deposits are found, we can suggest that

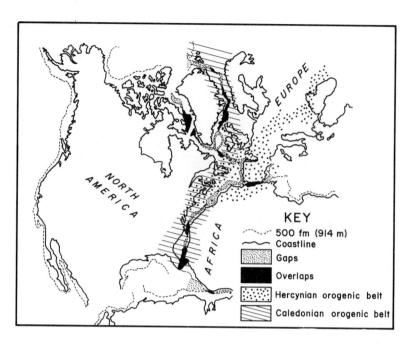

Figure 11.4 The fit of Europe with North America and Africa in terms of topography and orogenic belts.

those deposits formed in a low latitude at that time. The distribution of certain kinds of animals or plants may also be used to reveal climatic belts. Thus many corals live only in shallow equatorial waters.

Very many examples of ancient climatic belts now found displaced from their original position have been recognised. In the British Permian rocks, formed some 270 million years ago, there are cross-bedded sandstones representing hot desert conditions. Therefore since Permian times, Britain may have drifted closer to the north geographic pole by a distance comparable with 3000km. Another good example comes from Africa. In the Saharan region, which today is a hot desert, there are widespread Ordovician glacial deposits, 440 to 500 million years old (Figure 11.5). These are associated with glacially striated pavements. Thus in Ordovician times, North Africa lay at one of the geographical poles and was buried by an ice-sheet comparable in size with the modern ice-sheet of Antarctica. North Africa has moved at least 10 000km since Ordovician times!

The glacial deposits formed 250 to 300 million years ago in Permo-Carboniferous times give perhaps the most impressive climatic evidence favouring continental drift. These deposits occur in southern Africa, South America, Antarctica, Australia, and India and Ceylon (Figure 11.5). By matching their distributions in these areas, it is possible to reassemble a single large continent, called *Gondwanaland*, which seems to have lain over the south geographic pole. The pattern of glacial striae suggests that, as in Antarctica today, the

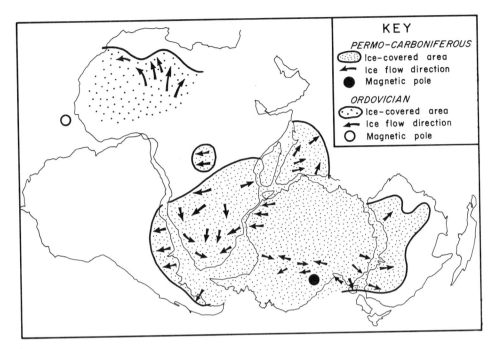

Figure 11.5 The Ordovician glaciation of North Africa and the Permo-Carboniferous glaciation of Gondwanaland in the southern hemisphere. A reconstruction based on the fit of continents, occurrence of glacial deposits and directions of glacial striae and other structures.

ice flowed from the centre towards the edges of the Gondwanaland ice-sheet. If this reconstruction is correct, then South America, Africa, Antarctica, Australia and India have drifted away from each other by thousands of kilometres since Permian times, in order to reach their present positions.

These conclusions have been independently tested in recent years by measuring the *ancient magnetic field* (*palaeomagnetism*) of the Earth. To see how this is possible, we should explain some features of the Earth's present magnetic field, the *geomagnetic field*.

The external geomagnetic field is very similar to that produced by a bar magnet or by a straight coil of wire through which an electrical current is passed (Figure 11.6). It is a *dipole field*, having two poles. Thus a *compass needle* turning in a *horizontal plane* on the Earth's surface has its north pole attracted to the north magnetic pole of the geomagnetic field. Now the north and south magnetic poles lie very close to the north and south geographic poles, through which passes the Earth's axis of rotation. The magnetic poles wander a little relative to the geographic poles, but never diverge from them by more than about 11° latitude, so that we may safely claim that on average the magnetic and geographic poles coincide in position. Another feature of the geomagnetic field is that, between the equator and the poles, the lines of

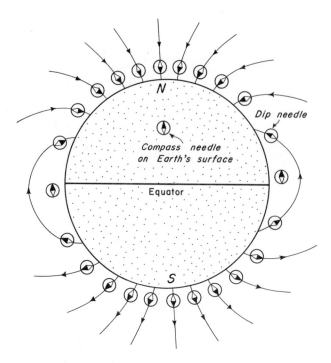

Figure 11.6 A model of the geomagnetic field, as shown by the behaviour of compass and dip needles.

magnetic force vary in their inclination to the surface of the Earth. This can be shown using a *dip needle*, a carefully balanced magnet free to turn in a *vertical plane*. The dip needle in the plane of a meridian is horizontal at the equator, but points vertically at the poles. Hence by taking measurements from a compass needle and a dip needle at any point on the surface of the Earth, we can say what is the latitude of that point.

When a rock is formed, as by the solidification of a magma, the iron minerals crystallising out become magnetised in the direction of the prevailing geomagnetic field. Consequently, the rock preserves a permanent record of the direction of the geomagnetic field which existed at the time and place where it formed. A specimen of the rock contains, so to speak, a *frozen compass and dip needle* representing the ancient geomagnetic field. Geologists have discovered how to measure the direction of the fossilised geomagnetic field, from a carefully collected specimen taken to the laboratory. From these measurements we can state the latitude of the place where each specimen came from at the time the rock was formed, and hence the positions of the ancient poles. This is done by referring directly to the relationship observed today between the direction of the geomagnetic field and latitude.

The palaeomagnetic results should agree with the evidence of the glacial

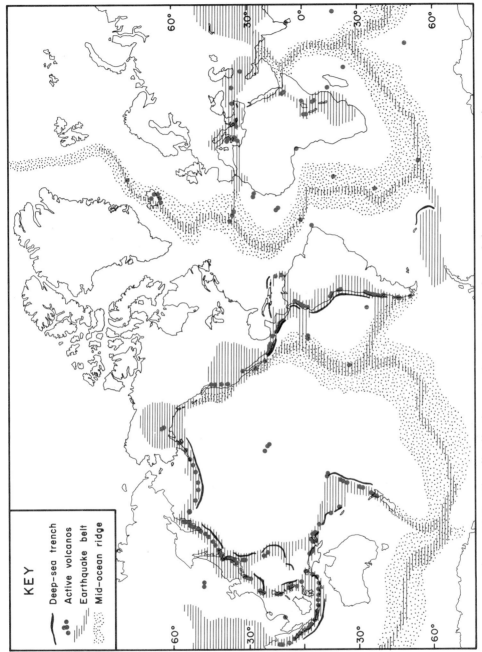

Figure 11.7 The distribution on the Earth's surface of deep-sea trenches, active volcanoes, and strong earthquakes.

deposits (Figure 11.5). The south magnetic pole in Ordovician times was calculated to be just off the west coast of Africa, close to the Ordovician glacial rocks. The south magnetic pole in Permo-Carboniferous times lay in Antarctica, within the region of Permo-Carboniferous glacial sediments. The agreement between the climatic and the magnetic evidence is therefore very good.

These lines of evidence make a strong case. There seeems little doubt that, in the geologically recent past, a large region of continental crust broke up into several smaller continents, which later drifted apart by distances measured in thousands of kilometres.

Sea-floor spreading and plate tectonics
Now that we can be reasonably certain that the continents have drifted, there are new questions to be asked. Do the continents drift like separate corks on water, or is each wandering continent a part of a larger piece of lithosphere in motion? The answer introduces the latest of all ideas concerning the Earth, namely, that the *sea floors are spreading* and that the entire lithosphere is divided up into a small number of *lithospheric plates* in relative motion. These plates are, of course, considered to be very large. Some of them appear large enough to carry a whole continent and a substantial part of an ocean.

These ideas arose in the minds of scientists puzzled by a number of major features of the Earth, and in particular by the *geographical coincidence* of several of those features (Figure 11.7). It was noted that the crust was thinner beneath the oceans than under the continents, as well as being different in composition. *Deep-sea trenches* are important features on the Earth's face, particular on the edges of the Pacific Ocean. Remarkably, this ocean is also bordered by large numbers of *active volcanoes,* as in the Andes, Alaska and Japan. Notice how the volcanoes form a ring which lies mainly outside the ring of trenches. However, volcanoes are also associated with the mid-ocean ridges, for instance, Hekla and others in Iceland, and there is a belt of them in the Mediterranean. The third remarkable coincidence is that many *strong earthquakes* occur beaneath the lands that border the Pacific Ocean, in roughly the same places as the trenches and volcanoes. However, frequent strong earthquakes occur also beaneath the crests of the mid-ocean ridges. Another important earthquake belt stretches from the Far East, through the Himalayan region, into the Mediterranean where there are also volcanoes. Notice the extent to which these belts of earthquakes and volcanoes coincide in position with the youngest fold mountains (Figure 1.4).

A fourth fact to be explained is the distribution of *earthquake foci* in depth, the focus of an earthquake being the point at which the shock waves originate. Where there is a deep-sea trench, the foci cluster near an imaginary plane which dips steeply downwards, under the adjacent continent, to a depth of about 500km (Figure 11.8). This concentration of earthquake foci is called a *Benioff zone,* after its discoverer. Now experience teaches us that the focus of an earthquake is located on a fault plane, and that the shock expresses fault movement. Does the existence of a Benioff zone under a continent bordered

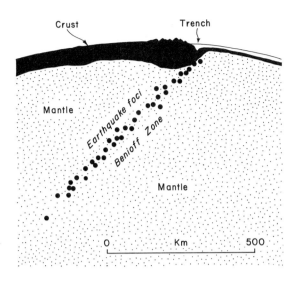

Figure 11.8 Steeply dipping zone of earthquake foci beneath a continent bordered by a deep-sea trench. No vertical exaggeration.

by a trench mean there is a gigantic fault there? If the Benioff zone does mean a zone of slipping, what parts of the Earth are in relative motion? We can ask another question. What is the meaning of the kind of continental margin found in the Atlantic Ocean, without either trench or Benioff zone?

We now think that these features, together with the basaltic composition of the oceanic crust, can be explained by the idea of *sea-floor spreading* (Figure 11.9). According to this idea, the mid-ocean ridges lie above the places where the *convection currents* within the middle portions of the Earth's mantle diverge, tending to drag parts of the lithosphere apart. Under the influence of these currents, new oceanic crust continuously appears along the mid-ocean ridges by the *injection and extrusion of magma* from the mantle, into and through the faults parallel to the crests of the ridges. In this way a *sheet of lithosphere* including oceanic crust is produced on which the lighter continents ride, like parcels on a conveyor belt. But because the Earth has a constant finite surface area, there must be places where the lithosphere created at the mid-ocean ridges is *thrust back* into the interior of the Earth. The Benioff zones are thought to indicate these places, where one portion of lithosphere is sliding underneath another portion. The pushing of one portion under another explains the trenches and the frequent strong earthquakes whose foci range between small and great depths. It also explains the association of *volcanoes* with *trenches* and *earthquakes*. The sliding must lead to huge amounts of frictional heat, causing the lithosphere near the Benioff zone to melt. The magma so produced would tend to rise and form volcanoes at the surface. Moreover, such volcanoes would be correctly placed in relation to the trenches, that is, in a ring outside them.

135

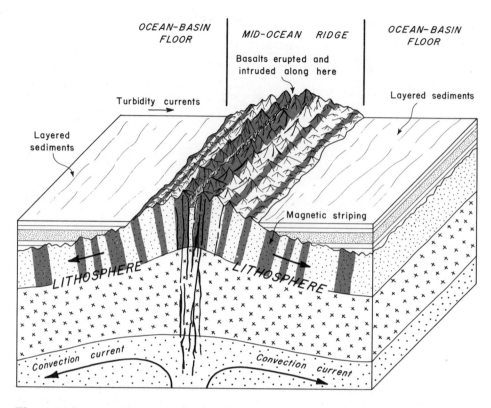

Figure 11.9 A model of sea-floor spreading by the injection of magma along the mid-ocean ridges under the influence of deep convection currents.

Thus we can divided the margins of continents into two types, *active* and *passive* (Figure 11.10). The active type is marked by a trench, a Benioff zone, and by volcanoes. One portion of the lithosphere is being pushed under another at such a margin, by the action of convection currents. The passive variety has no trench, no Benioff zone, and no volcanoes. A passive margin shows that the continent is riding along on a lithospheric 'conveyor belt' covered otherwise with oceanic crust, continent and ocean moving together and in the same direction.

What proof have we that the sea floor spreads as described? If spreading occurs in this way, the igneous rocks and the sedimentary deposits lying deepest below the ocean bed should become *progressively older* as we go *further away* on each side from a mid-ocean ridge (Figure 11.9). This prediction was recently shown to be true by studying rocks obtained from borings into the ocean bed beneath deep water. The oldest rocks found were early Jurassic, showing that the oceans as we see them today began to form in the early Mesozoic, some 180 million years ago.

The second line of evidence to support sea-floor spreading takes us again to the geomagnetic field. We have found by studying the magnetism of lavas that the geomagnetic field has *reversed in polarity many times* in the past few

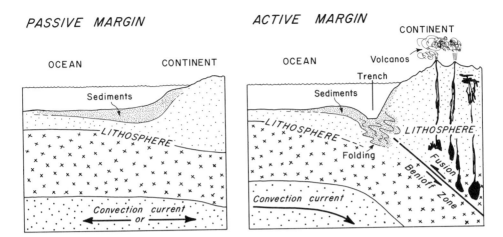

Figure 11.10 Cross-sectional vertical profiles comparing active and passive continental margins.

tens of millions of years. That is, the north and south magnetic poles have repeatedly changed places. By dating the same rocks it has been possible to say exactly when these changes or *reversals* in the geomagnetic field took place. Now because the oceanic crust is basalt, we should find evidence of these magnetic reversals from the ocean floor, if the spreading occurred as suggested. This evidence was obtained as a record of a pattern of stripes of normally and reversely magnetised rocks lying symmetrically about mid-ocean ridges (Fig. 11.9). We can even calculate the rate of spreading, because the date of each change in the magnetic field is known. In the Atlantic Ocean it is 1 to 2cm per year on each flank of the mid-ocean ridge, but in the Pacific Ocean is 3cm or more per year. These rates are astonishingly large and show that the Earth is changing rapidly.

Hence the earthquake belts and trenches define the edges of large portions of the lithosphere each of which behaves essentially as a rigid plate, a *lithospheric plate*. At the mid-ocean ridges fresh material is being added to the plates, whereas at the sites of the trenches the plates are being thrust back into the mantle. These plates are few in number but huge in extent. The Pacific Plate, for instance, consists of oceanic crust except for a small piece of continent represented by California west of the San Andreas Fault. The African Plate comprises Africa and Arabia and the surrounding Atlantic and Indian oceans up to the mid-ocean ridges. Then there are the American, Eurasian, Indian and Antarctic Plates, besides numerous smaller ones.

The association of young fold mountains with active continental margins (Figure 1.4) suggests that the *movement of the plates* may explain *orogenic fold-belts*. In and near the trenches, for example, one might expect to find deep-sea sediments that have become folded on account of the movement. An extensive fold-belt of sub-global proportions could easily be formed if two continents, carried on different plates, collided together. That fold-belt would be made up of a wide variety of sedimentary deposits but as well would include volcanic and intrusive igneous rocks. Although plate movement seems to account for the Alpine fold-belt, geologists are not yet agreed about the role of plate tectonics in the more distant past.

Index

Abrasion 105
Abyssal hills 89
Abyssal plains 16, 88, 89
Africa 76, 128
Alluvial fans 55
Americas 76, 128
Angle of rest 46, 59
Antarctica 37, 53, 92, 96, 101, 128
Aquicludes 39
Aquifers 39
Arctic Ocean 15, 76
Artesian basin 40
Atlantic Ocean 15, 81, 87, 88, 89, 91, 97, 99, 135, 137
Atmosphere 12, 37
Atmospheric circulation 30
Australia 53, 54, 128
Avalanches 47, 59

Backshore 72
Barrier islands 69–71
Basalt 15, 136, 137
Bays 68
Beaches 62, 63, 65, 71–73
Bedding 18
Bed load 25, 45
Benioff Zone 134, 136
Berm 72
Blakeney Spit 74
Bottom sampling 75–76
Breaker zone 72
Brittle substances 114
Brunizem 24

Capacity 34
Carbonation 21
Caves:
    limestone 21, 26, 40
    sea 68
Chalk 39, 40
Chernozem 24
Cirques 107
Cleavage 112, 122–124
Cleavage fan 124
Cliffs 29, 59, 65–69
Climatic change 49
Climatic zones 53, 129–131
Coastal plains of alluvium 51, 69
Coasts, agents affecting 63
Competence 34
Competent and incompetent beds 120–121
Continental crust 14
Continental drift 128–134

Continental margins 14, 87, 136
Continental rise 87
Continental shelves 74, 76, 80–85, 87
Continental slope 87
Continents, geological fit of 129
Continents, topographic fit of 128
Convection currents 135
Core 11
Corkscrew flow in river meanders 45–46
Corkscrew flow in the wind 60
Crevasses 109
Cross-bedding 47, 59, 60, 84
Cross-lamination 47
Crust of Earth 11, 14, 125
Crustal movements 12, 126
Currents, measurement of 76
Cut-off channels 47

Deep-sea fans 89
Deep-sea sediments 96–100
Deep-sea trenches 16, 87, 88, 134, 137
Deflation 57, 62
Deformation 12, 112
Deformed objects in rocks 119
Deltas 50, 69
Deposition 16, 25
Deserts:
    general 52
    drainage in 54
    Permian 130
    rocky 57
    sandy 57–62
    soils in 24
    stony 57
Dissolved load 26, 28, 30
Doldrums 53
Dorset coast 68, 74
Drifts 57, 59
Drowned valleys 49
Drumlins 108
Ductile substances 113
Dunes:
    ancient desert examples 129
    at sea-coast 62, 71
    barkhan 59
    longitudinal 60–61
    on outwash plains 110
    on river beds 47
    on sea bed 84
    star-shaped 61
    transverse 60
    wind-blown 48, 57

Dykes 39

Earthquakes 30, 94, 134
Echo-sounding 74
Eluviation 23
English Channel 15
Entrainment 25, 34, 36
Eskers 109
Europe 128
Evaporation 35, 38
Evapotranspiration 38

Face of the Earth 13, 14
Faults:
    general 39, 112
    normal 117
    reverse 117
    tear or wrench 117
Folds:
    general 112
    anticlinal 122
    basin-shaped 40
    dome-shaped 122, 124
    monoclinal 122, 124
    recumbent 122, 124
    synclinal 122
Fold-belts:
    general 112
    Alpine (Tertiary) 15, 112, 125, 138
    Caledonian (mid-Palaeozoic) 125
    explained by plate tectonics 137
    Hercynian (Permo-Carboniferous) 125
Force:
    gravitational 36, 112
    pressure 25
    shearing 25, 114
    tension or compression 114
Fracture zones 92
Frosted sand grains 57

Geocycle 13
Geological agents 13, 25, 34
Geomagnetic field 131–132, 136
Glacial abrasion 31, 105
Glacial striae 105, 108, 109, 111, 130
Glacial till 107
Glaciations:
    Ordovician 130
    Permo-Carboniferous 130–131
    Quaternary 100

Glacier erosion 104–107
Glaciers 25, 31, 38, 78, 92, 100–104
Glacio-marine sediments 96, 110–112
Gondwanaland 130
Grand Banks of Newfoundland 94
Gravel 25, 35, 47, 57, 71, 82, 83, 109
Great Glen Fault 117
Greenland 53
Grooving by ice-bergs 110
Ground moraine 108, 109
Groundwater 28, 38, 40
Gulf of Mexico 16, 51, 64, 80

Headlands 67
Hydrogenous sediments 96
Hydrographs 43
Hydrological cycle 37
Hydrolysis 20
Hydrosphere 12, 37

Ice 31, 101
Ice ages 100
Ice-bergs 78, 92, 110
Ice-caps 102
Ice-rafting 92, 110–112
Ice-sheets 38, 101, 128
Ice-shelves 102
Illuviation 23
Indian Ocean 15, 87, 88, 91, 97, 98
Internal drainage 54
Intertidal flats 71
Isostasy and isostatic equilibrium 126

Joints 18, 112, 114–116
Juvenile water 37

Kames 109
Karst 40, 42

Landslips 67
Lateritic soils 24
Lithosphere 12, 135
Lithospheric plates 137
Loess 110
Longshore currents 65, 72
Longshore transport 73

M discontinuity 11
Manganese nodules 95
Mantle 11
Meandering 43, 45–47
Meltwater 104, 109
Mesas 55
Meteoric water 37

Mid-ocean ridges 14, 86, 89–92, 135, 137
Mineral stability 19
Mississippi delta 50
Moraines 107–108
Mud (silt and clay) 26, 35, 48, 71, 81, 82, 84, 92, 93, 96
Mud cracks 48
Mudflows 67

Nepheloid layers 93
Niger delta 50, 71
Nigerian continental shelf 81
North Sea 16, 50, 71, 82–85

Ocean basins 14, 136
Ocean basin floors 14, 87, 89
Ocean ridges 87
Oceanic circulations 25, 30, 35, 85
Oceanic crust 15
Orford Ness 73
Outwash plains 109
Overland sheet flows 25
Ox-bow lakes 47
Oxidation 21

Pacific Ocean 15, 87, 88, 89, 91, 100, 134, 137
Palaeomagnetism 132–134
Parallel lamination 47, 70
Patagonian Desert 53
Pediments 55
Pelagic sediments 96
Piedmont glaciers 102
Plankton 92, 94–95
Plate tectonics 134–138
Podzol 23
Point bars 43
Proglacial deposits 109–110
Promontories 68

Quarrying 104

Radioactive carbon 80
Raised beaches 128
Regression 79
Rift valleys 92, 117
Rip currents 65
Ripples:
  on river beds 47
  wave-formed 70, 77
  wind-blown 57, 59
Rivers:
  general 25, 73, 79
  braided 43, 109
  deposits of 45–51
  discharge 43
  drainage basin and net 42
  floodplains 45
  levees 47

meandering 43, 45
terraces 48–49
underground 25, 26
Roches moutonées 105

Sahara 37, 52, 60, 92
Salt marshes 71
Saltation 30, 57
San Andreas Fault 117
Sand 25, 35, 47, 57, 69–70, 71, 81, 82, 84, 109, 110, 129
Sand ribbons 84
Sand seas 57–62
Scandinavia 108, 109, 128
Sea-floor spreading 134–138
Sea-level change 49, 78–80
Sea-stacks 68
Seismic profiler 85
Shelf sediments 80–85
Shelf-edge 74, 93
Side-scan sonar 75
Sills 39
Sink-holes 41
Slates 124
Slides 30, 94
Slumps 30
Soil:
  general 25, 38, 79
  flowing 25, 32–33
  horizons 21
  types and distribution of 23–24
Solution 40
Spits 73–74
Springs 39
Stalactites 41
Stalagmites 41
Stone fabrics in tills 108, 109
Storms 30, 78, 80, 92, 93
Storm tides 78, 80
Strain 113
Strain markers 118–120
Structure of the Earth 11
Submarine canyons 87, 88, 93
Submarine rises 89
Suspended load 26, 31
Swamps 47
Swash and backwash 72

Takla Makan Desert 54
Tectonic structures 112
Terminal moraines 108
Terrigenous sediments 96
Tidal channels 71
Tidal currents 29, 63, 78, 82
Tidal deltas 71
Tidal range 64
Tides 25, 29, 63
Tombolos 74
Trade winds 53, 92
Transgression 79, 81, 82

Transport 25, 34
Turbidites 96
Turbidity currents 25, 29, 93–94
Turbulence 26

Underwater photography 85, 97–99
U-shaped valleys 105

Valley glaciers 102
Valley trains 109

Ventifacts 57, 110
Volcanoes 134

Wadis 55
Water table 38
Wave-base 77
Waves 25, 28, 35, 64, 82
Wave currents 28, 76–77
Wave refraction 73
Weathering:
  general 13, 15, 17
  chemical 19
  mechanical 17
  onion-skin 20
  role of organisms in 17, 18, 21
Wells 39, 40
Wind 25, 35, 63, 70, 110
Wind-blown dust 92

Yardangs 57
Yield point 113

Zonal winds 53